Flow Cytometry

First Principles

Second Edition

Flow Cytometry

First Principles

Second Edition

Alice Longobardi Givan

The Herbert C. Englert Cell Analysis Laboratory
of the Norris Cotton Cancer Center
and Department of Physiology
Dartmouth Medical School
Lebanon, New Hampshire

WILEY-LISS

A John Wiley & Sons, Inc., Publication

New York • Chichester • Weinheim • Brisbane • Singapore • Toronto

This book is printed on acid-free paper. ⊚

Copyright © 2001 by Wiley-Liss, Inc. All rights reserved.

Published simultaneously in Canada.

For ordering and customer service, call 1-800-CALL-WILEY.

Library of Congress Cataloging-in-Publication Data:

Givan, Alice Longobardi.
  Flow cytometry : first principles / Alice Longobardi Givan.
    p. cm.
  Includes bibliographical references and index.
  ISBN 0-471-38224-8
  1. Flow cytometry. I. Title.
  [DNLM: 1. Flow Cytometry—methods.   QH585.5.F56 G539f]
QH585.5.F56G58   1992
574.87′028—dc20
DNLM/DLC
for Library of Congress   92-5004

10 9

*The First Edition of this book was dedicated to my parents,
Violet Litwin Longobardi and Vincent Longobardi, Jr.,
with gratitude for the example they set,
with pride in their achievements,
and with love.*

*The Second Edition is dedicated to
Curt, Ben, and Becky
(not only because I promised them that their turn would come).*

Contents

Preface

Although flow cytometry is simply a technique that is useful in certain fields of scientific endeavor, there is, at the same time, something special about it. Few other techniques involve specialists from so many different backgrounds. Anyone working with flow systems for any length of time will realize that computer buffs, electronics experts, mathematicians, optical and fluidics engineers, and organic chemists rub shoulders with biologists, physicians, and surgeons around the flow cytometer bench.

And it is not just a casual rubbing of shoulders, in passing, so to speak. Many of the specialists involved in flow cytometry might, if asked, call themselves *flow cytometrists* because the second aspect of flow cytometry that distinguishes it from many other techniques is that flow cytometry has itself become a "field." Indeed, it is a field of endeavor and of expertise that has captured the imaginations of many people. As a result, there exists a spirit of camaraderie; flow cytometry societies, groups, meetings, networks, websites, journals, courses, and books abound.

A third aspect of flow cytometry (known sometimes simply with the acronym for fluorescence-activated cell sorter, *FACS*, or even more familiarly as just *flow*) that distinguishes it from many other techniques is the way in which its wide and increasing usefulness has continued to surprise even those who consider themselves experts. What began as a clever technique for looking at a very limited range of problems is now being used in universities, in hospitals, within industry, at marine stations, on submersible buoys, and on board ships; plans have existed for use on board space ships as well. The applications of flow cytometry have proliferated (and continue to proliferate) rapidly both in the direction of theoretical science, with

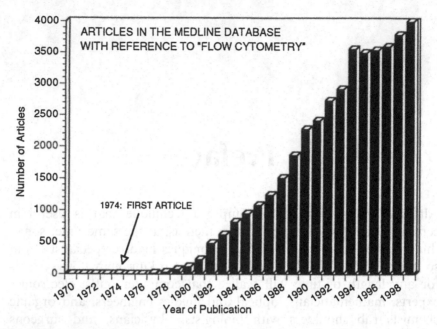

Fig. 1. Increasing reference to "flow cytometry" in the medical literature over the past three decades. The development of flow cytometers antedates the use of the term itself.

botany, molecular biology, embryology, biochemistry, marine ecology, genetics, microbiology, and immunology, for example, all represented; and in the direction of clinical diagnosis and medical practice, with hematology, bacteriology, pathology, oncology, obstetrics, and surgery involved. We are, at present, living through what appears to be a rapid phase in flow cytometry's growth curve (see Fig. 1).

Because flow cytometry is an unusual field, bringing together people with differing scientific backgrounds at meetings, on editorial boards, in hospital wards, on advisory panels, and at laboratory benches and reaching increasing numbers of workers in new and unpredicted areas of endeavor, there is, as a result, a need to provide both recent and potential entrants into this diverse community with a common basis of knowledge—so that we can all understand the vocabulary, the assumptions, the strengths, and the weaknesses of the technology involved. I have for many years taught new and future users of flow cytometers. My teaching attempts to present enough technical background to enable students, scientists, technologists, and clinicians to read the literature critically, to evaluate the benefits of the technique realistically, and, if tempted, to design effective protocols and interpret the results. I try to describe the theory of flow cytometry in a way that also provides a firm (and accurate) foundation for those few who will go on to study the technique in greater depth. Details of protocols are avoided, but my teaching attempts to give enough information about applications to provide concrete examples of general concepts and to allow some appreciation of the range of practical goals that the instruments are able to achieve. With some expansion (but with little change in style or objectives), my classroom and workshop teaching is the basis for this book.

Notes on the Second Edition: This new edition, while similar to the first edition in style and scope, has been modified in many ways. The arrangement of some material has been altered to present, in my opinion, a more coherent pedagogical sequence, reflecting my changing thoughts on teaching. While all chapters have been re-written to a significant extent, there are also some major expansions reflecting the progress of the field. In particular, I have included more detail about the cell as it passes through the laser beam; the laser/fluorochrome chapter has been expanded to include recently developed fluorochromes and multilaser options; a new chapter has been added on cytoplasmic staining; a discussion of apoptosis has been added to the chapter on DNA; the section on sorting has been expanded to a full chapter and includes high-speed sorting and alternative sorting methods as well as traditional technology; the clinical and research chapters have been updated and expanded considerably; the chapter on general references includes many of the recent excellent books in the field; and the chapter on the future of flow cytometry is now a subjective glimpse into the new decade from the vantage point of the year 2000.

Acknowledgments
(First Edition)

Realizing how much I have learned and continue to learn from others, I hesitate to single out a few names for particular mention. However, with the disclaimer that any list of people to whom I am indebted is not meant to be and, indeed, could never be complete, I must thank here the following friends, mentors, and colleagues who have had a very direct impact on the writing of this book: George Proud, for having had enough insight into the importance of a flow cytometer for transplantation surgery to want to have one; Ivan Johnston and Ross Taylor, for hospitality (and the Harker Bequest Fund, the Northern Counties Kidney Research Fund, and the Newcastle Health Authority, for financial assistance) within the Department of Surgery at Newcastle University; Brian Shenton, for introducing me to the field of flow cytometry and to the joys and hazards of clinical research; Mike White, for bailing me out (often figuratively and once or twice literally) on so many occasions; Paul Dunnigan, for teaching me about lasers (and also about fuses, relays, and loose wires); Ian Brotherick, for the animal amplification figure, and both he and Alison Mitcheson for good humor in the lab beyond all reasonable expectation; Terry Godley, for being an extremely good and (more importantly) a very communicative flow cytometry service engineer; Ray Joyce, for considerable assistance with the design and drawing of many of the diagrams; the scientists and clinicians at Newcastle University and Durham University (many acknowledged in the figure legends), for providing me with a continuous source of interesting and well-prepared cell material on which to practice my flow technique; Paul Guyre, for giving me a supportive and remarkably enjoyable re-introduction to science in the New World; Daryll Green, for giving me the benefit of his expertise in both physics and flow cytometry by reading and commenting about the chapters on instrumentation and information; Brian Crawford, for being a literate and encouraging editor in the face of my rank inexperience; the late Robert L. Conner, for providing me with that critical first of my still-continuing scientific apprenticeships; Curt Givan, for his unfailing loyalty and for his skill in reading the entire manuscript with two eyes—one eye that of an old-fashioned grammarian who abhor dangling participles and the other that of a modern scientist who knows nothing about flow cytometry; and Ben Givan, for the two drawings in Figure 8.1 [Fig. 10.1 in the Second Edition], Becky Givan, for organizational magic when it was very badly needed, and both kids for *lots* of encouragement and some pretty funny suggestions for a title.

A.L.G., 1991, Newcastle upon Tyne.

Acknowledgments
(Second Edition)

I continue to be immensely grateful to the people whom I acknowledged in the First Edition of FCFP; their contributions are still current and form the foundation for this new edition. I have, however, during preparation of the new edition, received considerable additional help. During the 9 years since the publication of the first edition of this book, changes have taken place in my life as well as in the field of flow cytometry. Now working at Dartmouth Medical School, I have learned much about clinical cytometry, flow analysis, and computers from Marc Langweiler; Marc taught me the acronym "RTFM," but has been restrained at using it in response to my naïve questions and I am grateful. In addition, Marc was a particular help to me with suggestions for Chapter 10. Gary Ward, the flow sorting supervisor in the Englert Cell Analysis Laboratory at Dartmouth, has, by his competence, made my sorting skills rusty, but he has allowed me to pretend that I run "his" sorter when I need a dramatic photograph, and I am grateful for his good humor. I am also grateful to Ken Orndorff, who supervises the imaging service in the Englert Lab; he has been admirably tolerant of flow cytometrists (including me) who have slowly begun to realize that a dot on a dot plot may not be enough and that they need to know what their cells look like. In addition, I want to thank the many users of the instruments in the Englert Lab for their ability to provide me with continuing challenges.

First Colette Bean and then Luna Han at Wiley inherited editorial supervision of this book from Brian Crawford and I owe them considerable thanks for their persistence and patience. At Dartmouth, I am grateful for secretarial assistance from Mary Durand and artistic help from Joan Thomson. Dean Gonzalez at Wiley, with artistry and stamina, is responsible for the great improvement in figure clarity between the two editions of this book. I thank him.

During these last 9 years, I have had the opportunity to expand my acquaintance with members of the greater flow community. Among many from whom I have learned much, I would like especially to mention a few: Louis Kamentsky shared with me his lecture notes for a wonderful keynote address he gave at Dartmouth on the history of flow cytometry, and I am grateful to him. Howard Shapiro has been a source of limitless cytometric information cloaked seductively in outrageous humor; I thank him for his knowledge, for his humor, and for his generosity. Paul Robinson has made many contributions to flow cytometry, but I want particularly to mention his organization of the Purdue University flow network; I am grateful to him for the information, the moral support, and the distance friendships that the network has given me. Over the years, I have had the privilege of teaching workshops and courses with a group of people who are devoted not just to flow technology (although they are talented flow cytometrists) but also to education of students in the correct use of the technique. Indeed, I also detect their devotion to the use of flow cytometry as a platform for teaching a rigorous approach to science in general. In thanks for all I have learned from these workshops, I wish, particularly, to mention here Bruce Bagwell, Ben Hunsberger, James Jett, Kathy Muirhead, Carleton Stewart, Joe Trotter, and Paul Wallace.

And—as ever—Curt Givan came through with unflappable proofreading skills and many cups of tea.

A.L.G., 2000, Lebanon, New Hampshire.

1

The Past as Prologue

Flow cytometry, like most scientific developments, has roots firmly grounded in history. In particular, flow technology finds intellectual antecedents in microscopy, in blood cell counting instruments, and in the ink jet technology that was, in the 1960s, being developed for computer printers. It was the coming together of these three strands of endeavor that provided the basis for the development of the first flow cytometers. Because thorough accounts of the history of flow cytometry have been written elsewhere (and make a fascinating story for those interested in the history of science), I cover past history here in just enough detail to give readers a perspective as to why current instruments have developed as they have.

Microscopes have, since the seventeenth century, been used to examine cells and tissue sections. Particularly since the end of the nineteenth century, stains have been developed that make various cellular constituents visible; in the 1940s and 1950s, fluorescence microscopy began to be used in conjunction with fluorescent stains for nucleic acids in order to detect malignant cells. With the advent of antibody technology and the work of Albert Coons in linking antibodies with fluorescent tags, the use of fluorescent stains gained wider and more specific applications. In particular, cell suspensions or tissue sections are now routinely stained with antibodies specific for antigenic markers of cell type or function. The antibodies are either directly or indirectly conjugated to fluorescent molecules (most usually fluorescein or rhodamine). The cellular material can then be examined on a glass slide under a microscope fitted with an appropriate lamp and filters so that the fluorescence of the cells can be excited and observed (Fig. 1.1). The fluorescence microscope allows us to see cells, to identify them in terms of both their physical structure and

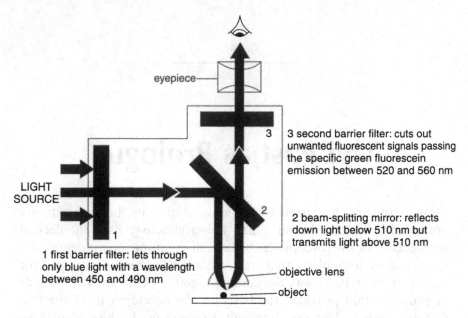

Fig. 1.1. The optical path of a fluorescence microscope. In this example, the filters and mirrors are set for detection of fluorescein fluorescence. From Alberts et al. (1989).

their orientation within tissues, and then to determine whether and in what pattern they fluoresce when stained with one or another of the specific stains available. In addition, a microscope can also be fitted with a camera or photodetector, which will then record the intensity of fluorescence arising from the field in view. The logical extension of this technique is image analysis cytometry, digitizing the output to allow precise quantitation of fluorescence intensity patterns in detail (pixel by pixel) within that field of view. The development of monoclonal antibody technology (for which Köhler and Milstein were awarded the Nobel Prize) led to a vast increase in the number of cellular components that can be specifically stained and that can be used to classify cells. Whereas monoclonal antibody techniques are not directly related to the development of flow technology, their invention was a serendipitous event that had great impact on the potential utility of flow cytometric systems.

In 1934, Andrew Moldavan in Montreal took a first step from static microscopy toward a flowing system. He suggested the development of an apparatus to count red blood cells and neutral-red–stained yeast cells as they were forced through a capillary on a microscope stage. A photodetector attached to the microscope eyepiece would register each passing cell. Although it is unclear from Moldavan's paper whether he actually ever built this cytometer, the development of staining procedures over the next 30 years made it obvious that the technique he suggested could be useful not simply for counting the number of cells but also for quantitating their characteristics.

In the mid-1960s, Louis Kamentsky took his background in optical character recognition and applied it to the problem of automated cervical cytology screening. He developed a microscope-based spectrophotometer (on the pattern of the one suggested by Moldavan) that measured and recorded ultraviolet absorption and the scatter of blue light ("as an alternative to mimicking the complex scanning methods of the human microscopist") from cells flowing "at rates exceeding 500 cells per second" past a microscope objective. Then, in 1967, Kamentsky and Melamed elaborated this design into a sorting instrument (Fig. 1.2) that provided for the electronic actuation of a syringe to pull cells with high absorption/scatter ratios out of the flow stream. These "suspicious" cells could then be subjected to detailed microscopic analysis. In 1969, Dittrich and Göhde in Münster, Germany, described a flow chamber for a microscope-based system whereby fluorescence intensity histograms could be generated based on the ethidium bromide fluorescence from the DNA of alcohol-fixed cells.

During this period of advances in flow microscopy, so-called Coulter technology had been developed by Wallace Coulter for analysis of blood cells. In the 1950s, instruments were produced that counted cells as they flowed in a liquid stream; analysis was based on the amount by which the cells increased electrical resistance as they displaced isotonic saline solution while flowing through an orifice. Cells were thereby classified more or less on the basis of their volume because larger cells have greater electrical resistance. These Coulter counters soon became essential equipment in hospital hematology laboratories, allowing the rapid and automated counting of white and red blood cells. They actually incorporated many of the features of analysis that we now think of as being typical of flow cytometry: the rapid flow of single cells in file through an orifice, the electronic detection of signals from those cells, and the automated analysis of those signals.

At the same time as Kamentsky's work on cervical screening,

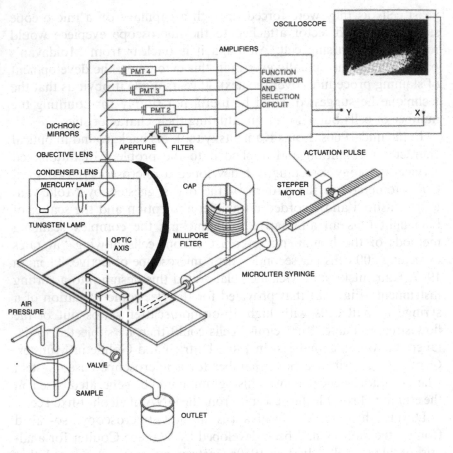

Fig. 1.2. A diagram of Kamentsky's original flow sorter. From Kamentsky and Melamed (1967). Science 156:1364–1365. Copyright AAAS.

Mack Fulwyler at the Los Alamos Laboratory in New Mexico had decided to investigate a problem well known to everyone looking at red blood cells in Coulter counters. Red cells were known to show a bimodal distribution of their electrical resistance ("Coulter volume"). Anyone looking at erythrocytes under the microscope cannot help but be impressed by the remarkable structural uniformity of these cells; Fulwyler wondered if the bimodal Coulter volume distribution represented differences between two classes of these apparently very uniform cells or, alternatively, whether the bimodal profile was simply an artifact based on some quirky aspect of the electronic resistance measurements. The most direct way of testing these two alternatives was to separate erythrocytes according to their electronic resistance signals and then to determine whether the separated classes remained distinct when they were re-analyzed.

The technique that Fulwyler developed for sorting the erythrocytes combined Coulter methodology with the ink jet technology being developed at Stanford University by RG Sweet for running computer printers. Ink jet technology involves the vibration of a nozzle so as to generate a stream that breaks up into discrete drops and then the charging and grounding of that stream at appropriate times so as to leave indicated drops, as they break off, carrying an electrical charge. For purposes of printing, those charged drops of ink can then be deflected to positions on the paper as required by the computer print messages. Fulwyler took the intellectual leap of combining this methodology with Coulter flow technology; he developed an instrument that would charge drops containing suspended cells, thereby allowing deflection of the cells (within the drops) as dictated by signals based on the cell's measured Coulter volume.

The data from this limited but pioneering experiment led to a conclusion that with hindsight seems obvious: Erythrocytes are indeed uniform. When red cells are sorted according to their electrical resistance, the resulting cells from one class or the other still show a bimodal distribution when re-analyzed for their electrical resistance profile. The bimodal "volume" signal from erythrocytes was therefore artifactual—resulting in part from the discoid (nonspherical) shape of the cells. The technology developed for this landmark experiment is the essence of all the technology required for flow sorting as we now know it. That experiment also, unwittingly, emphasized an aspect of flow cytometry that has remained with us to this day: Flow cytometrists still need to be continually vigilant (and know how to use a microscope) because signals from cells (particularly signals that are related to cell volume) are subject to artifactual influences and may not be what they seem. (Fulwyler's 1965 paper actually describes the separation of mouse from human erythrocytes and the separation of a large component from a population of mouse lymphoma cells; the experiments on the bimodal signals from red cells have been relegated to flow folk history.)

In 1953, PJ Crosland-Taylor, working at the Middlesex Hospital in London, noted that attempts to count small particles suspended in fluid flowing through a tube had not hitherto been very successful. With particles such as red blood cells, the experimenter must choose between a wide tube that allows particles to pass two or more abreast across a particular section or a narrow tube that makes microscopical observation of the contents of the tube difficult due to the different refractive indices of the tube and the suspending fluid. In addition, narrow tubes tend to block easily. In response to this dilemma, Crosland-Taylor applied the principles of laminar flow to the design of a flow system. A suspension of red blood cells was injected into the center of a faster flowing stream, thus allowing the cells to be aligned in a narrow central file within the core of the wider stream preparatory to electronic counting. This principle of hydrodynamic focusing was pivotal for the further development of the field.

In 1969, Marvin Van Dilla and other members of the Los Alamos Laboratory group reported development of the first fluorescence-detection cytometer that utilized the principle of hydrodynamic focusing and, unlike the microscope-based systems, had the axes of flow, illumination, and detection all orthogonal to each other; it also used an argon laser as a light source (Fig. 1.3). Indeed, the configuration of this instrument provided a framework that could support both the illumination and detection electronics of Kamentsky's device as well as the rapid flow and vibrating fluid jet of Fulwyler's sorter. In the initial report, the instrument was used for the detection of fluorescence from the Feulgen-DNA staining of Chinese hamster ovary

Fig. 1.3. Marvin Van Dilla and the Livermore flow sorter in 1973. Photograph courtesy of the Lawrence Livermore National Laboratory.

cells and leukocytes as well as of their Coulter volume; however, the authors "anticipated that extension of this method is possible and of potential value." Indeed, shortly thereafter, the Herzenberg group at Stanford published a paper demonstrating the use of a similar cytometer to sort mouse spleen and Chinese hamster ovary cells on the basis of their fluorescence due to accumulation of fluorescein diacetate. These instruments thus led to systems for combining multiparameter fluorescence, light scatter, and "Coulter volume" detection with cell sorting.

These sorting cytometers began to be used to look at ways of distinguishing and separating white blood cells. By the end of the 1960s, they were able to sort lymphocytes and granulocytes into highly purified states. The remaining history of flow cytometry involves the elaboration of this technology, the exploitation of flow cytometers for varied applications, and the collaboration between scientists and industry for the commercial production of cytometers as user-friendly tools (Fig. 1.4). At the same time that these instruments began to be seen as commercially marketable objects, research and development continued especially at the United States National Laboratories at Los Alamos (New Mexico) and Livermore (California), but also at smaller centers around the world. At these centers, homemade in-

Fig. 1.4. Bernard Shoor (left) and Leonard Herzenberg at Stanford University with one of the original Becton Dickinson flow cytometers as it was packed for shipment to the Smithsonian Museum. Photograph by Edward Souza, courtesy of the Stanford News Service.

Fig. 1.5. Three user-friendly benchtop cytometers (in alphabetical order). **Top:** A Beckman Coulter ®XL™. **Middle:** A Becton Dickinson FACSCalibur. **Bottom:** A Dako Galaxy, manufactured for Dako by Partec.

struments continue to indicate the leading edge of flow technology. At the present time, this technology is moving simultaneously in two directions: On the one hand, increasingly sophisticated instruments are being developed that can measure and analyze more aspects of more varied types of particles more and more sensitively and that can sort particles on the basis of these aspects at faster and faster rates. On the other hand, a different type of sophistication has streamlined instruments (Fig. 1.5) so that they have become user-friendly and essential equipment for many laboratory benches.

FURTHER READING

Throughout this book, the "Further Reading" references at the end of each chapter, while not exhaustive, are intended to point the way into the specific literature related to the chapter in question. At the end of the book, "General References" will direct readers to globally useful literature. Titles in bold at the end of each chapter are texts that are fully cited in the General References at the end.

Coulter WH (1956). High speed automatic blood cell counter and size analyzer. Proc. Natl. Electronics Conf. 12:1034–1040.

Crosland-Taylor PJ (1953). A device for counting small particles suspended in a fluid through a tube. Nature 171:37–38.

Dittrich W, Göhde W (1969). Impulsfluorometrie dei Einzelzellen in Suspensionen. Z. Naturforsch. 24b:360–361.

Fulwyler MJ (1965). Electronic separation of biological cells by volume. Science 150:910–911.

Herzenberg LA, Sweet RG, Herzenberg LA (1976). Fluorescence-activated cell sorting. Sci. Am. 234:108–115.

Hulett HR, Bonner WA, Barrett J, Herzenberg LA (1969). Cell sorting: Automated separation of mammalian cells as a function of intracellular fluorescence. Science 166:747–749.

Kamentsky LA, Melamed MR (1967). Spectrophotometric cell sorter. Science 156:1364–1365.

Kamentsky LA, Melamed MR, Derman H (1965). Spectrophotometer: New instrument for ultrarapid cell analysis. Science 150:630–631.

Moldavan A (1934). Photo-electric technique for the counting of microscopical cells. Science 80:188–189.

Van Dilla MA, Trujillo TT, Mullaney PF, Coulter JR (1969). Cell microfluorimetry: A method for rapid fluorescence measurement. Science 163:1213–1214.

Chapter 1 in **Melamed et al.**, Chapter 3 in **Shapiro**, and Chapter 1 in **Darzynkiewicz** are good historical reviews of flow cytometry. Alberto Cambrosio and Peter Keating (2000) have used flow cytometry as a model for looking at historical changes in the way scientists use instrumentation to view the world: Of lymphocytes and pixels: The techno-visual production of cell populations. Studies in History and Philosophy of Biological and Biomedical Sciences 31:233–270.

2

Setting the Scene

As mentioned in the previous chapter, flow cytometry has been moving in two directions at once. The earliest flow cytometers were either homemade Rube Goldberg (Heath Robinson, U.K.) monsters or, a few years later, equally complex, unwieldy commercial instruments. These were expensive; they were also unstable and therefore difficult to operate and maintain. For these reasons, the cytometer tended to collect around itself the trappings of what might be called a flow facility—that is, a group of scientists, technicians, students, and administrators, as well as a collection of computers and printers, that all revolved around the flow cytometer at the hub. If a scientist or clinician wanted the use of a flow cytometer to provide some required information, he or she would come to the flow facility, discuss the experimental requirements, make a booking, and then return with the prepared samples at the allotted time. The samples would then be run through the cytometer by a dedicated and knowledgeable operator. Finally, depending on the operator's assessment of the skill of the end-user, a number, a computer print-out, or a computer disk would be handed over for analysis.

Just at the moment when such flow facilities were beginning to see themselves as essential components of modern research, the technology of flow cytometry began to move in a new direction. Users and manufacturers both began to realize that different laboratories have different instrumentation needs. A sophisticated sorter might be necessary for certain applications, but its demands on daily alignment and skilled maintenance are time-consuming and its research capabilities might be superfluous for routine processing of, for example, clinical samples. Instead of simply continuing to become larger, more expensive, more complicated, and more powerful, some flow cytometers started to become, in the late 1980s, smaller, less expensive, more accessible, and, although less flexible, remarkably stable. These instruments are called "benchtop" cytometers because they are self-contained and can be taken from their shipping crates, placed on a lab bench, plugged in, and (with luck) are ready to go.

The question arises as to how much training is required for use of these benchtop instruments. Pushing the buttons has become easy, but my prejudice on this issue should, of course, be obvious. If I believed that flow experiments could be designed and flow data could be acquired and analyzed appropriately by people with no awareness of the limitations and assumptions inherent in the technique, I would not have written this book. The actual operation of these small cytometers has been vastly simplified compared with that of the original research instruments. It certainly can be said that the new wave of cytometers has made flow analysis a great deal more accessible to the nonspecialist. A serious concern, however, is that the superficial simplicity of the instruments may lull users into a false sense of security about the ease of interpretation of the results. The basis for this concern is particularly clear in the medical community, where clinicians have been conditioned to expect that laboratory reports will contain unambiguous numbers; they may not be accustomed to the need for an intellectual framework in which to interpret those numbers. Anyone designing flow experiments or interpreting flow data needs some essential training in the technique; more training is required as the benchtop instruments (and the possible experiments) become more complex. At the high end, the operation of sorting instruments is best left to a dedicated operator; but, even here, the grass-roots user needs to understand flow cytometry in order to design an effective sorting protocol and communicate this protocol to the operator.

Although benchtop cytometers are less expensive than state-of-the-art instrumentation, they are still expensive. Therefore core facilities with shared instrumentation still provide for much of the current flow cytometric analysis. These shared facilities reflect a need by many for flow cytometric instrumentation, but also recognition of its high cost, its requirement for skilled maintenance and operation, and the fact that many users from many departments may each require less than full-time access. Such centralized facilities may have more than one cytometer. The trend now is to have one or more sophisticated instruments for specialized procedures accompanied by several benchtop cytometers as routine work horses. There are usually dedicated operators who run the sophisticated instruments, supervise use of the benchtop instruments, and ensure that all instruments are well-tuned for optimum performance. In addition, there may be a network of computers so that flow data can be analyzed and re-analyzed at leisure—away from the cytometer and possibly in the scientist's own laboratory. A cytometry facility may also provide centrifuges, cell incubators, and fluorescence microscopes to aid in cell preparation. Increasingly, such institutional flow facilities are also becoming general cytometry facilities; the use of imaging microscopes for cell analysis has proved to be a technique that complements flow work. Similar stains are used in both flow and image analysis systems. Flow cytometry is based on analysis of light from a large number of cells as they individually pass a detector. Image analysis studies the distribution of light signals emanating from a large number of individual positions scanned over a single stationary cell.

The funding of such central facilities often involves a combination of institutional support, direct research grant support, and charges to users (or patients). The charges to users may vary from nominal to prohibitive and may need to support everything from consumables like computer paper and buffers to the major costs of laser replacement, service contracts, and staff salaries. With the increasing complexity and variety of instrumentation and with the need for data organization, booking systems, and financial accounting, the running of these flow facilities can become an administrative task in itself. Therefore a "corporate identity" often evolves, with logos designed, meetings organized, newsletters written, and training sessions provided.

Moreover, a funny thing has been happening. What originally appeared to be a rift in the field of flow cytometry, isolating the "high tech" from the "benchtop" users, has turned out to be, instead, a continuum. State-of-the-art instruments continue to improve, marking the advancing edge of technological and methodological development. However, benchtop flow cytometers have followed along behind, becoming increasingly powerful and combining the advantages of stable optics and fluidics with the new capabilities demonstrated on last year's research instruments.

In this book, emphasis will be on the general principles of cytometry that apply equally to benchtop and state-of-the-art instru-

mentation. The chapter on sorting will, however, provide some
insight into the ability of research cytometers to separate cells. In
addition, the chapter on research frontiers provides a glimpse into the
advanced capabilities of today's research instruments (and, without
doubt, many of these capabilities will soon appear on tomorrow's
benchtop cytometers).

3

Instrumentation: Into the Black Box

I want first to clear up some confusion that results simply from words. A flow cytometer, despite its name, does not necessarily deal with cells; it deals with cells quite often, but it can also deal with chromosomes or molecules or latex beads or with many other particles that can be suspended in a fluid. Although early flow cytometers were developed with the purpose of sorting particles, many cytometers today are in fact not capable of sorting. To add to the semantic confusion, the acronym FACS is even applied to some of these non-sorting cytometers. Flow cytometry might be broadly defined as a system for measuring and then analyzing the signals that result as particles flow in a liquid stream through a beam of light. Flow cytometry is, however, a changing technology. Defining it is something like capturing a greased pig; the more tightly it is grasped, the more likely it is to wriggle free. In this chapter, I describe the components that make up a flow cytometric system in such a way that we may not need a definition, but will know one when we see it.

The common elements in all flow cytometers are

- A light source with a means to focus that light
- Fluid lines and controls to direct a liquid stream containing particles through the focused light beam
- An electronic network for detecting the light signals coming from the particles as they pass through the light beam and then converting the signals to numbers that are proportional to light intensity

· A computer for recording the numbers derived from the electronic detectors and then analyzing them

In this chapter I describe the way a flow cytometer shapes the light beam from a laser to illuminate cells; the fluid system that brings the cells into the light beam; and the electronic network for detecting and processing the signals coming from the cells. The next chapter will cover computing and analysis strategies for converting flow cytometric data into useful information.

ILLUMINATION OF THE STREAM

A laser beam has a circular, radially symmetrical cross-sectional profile, with a diameter of approximately 1–2 mm. Lenses in the cytometer itself are used to shape the laser beam and to focus it to a smaller diameter as it illuminates the cells. Simple spherical lenses can provide a round spot about 60 μm in diameter, with its most intense region in the center and its brightness decreasing out toward the edge of the spot. The decrease in intensity follows a Gaussian profile; cells passing through the middle of the beam are much more brightly illuminated than those passing near the periphery. By convention, the nominal diameter of a Gaussian beam of light refers to the distance across the center of the beam at which the intensity drops to 13.5% of its maximal, central value. Therefore, if a cell is 30 μm from the center of a nominal 60 μm beam, it will receive only 13.5% of the light it would receive at the center of that beam. At 10 μm from the center, the intensity is about 78% and, at 3 μm from the center, the intensity is 98% of the central intensity. Because intensity falls off rapidly at even small distances from the center of the beam, cells need to be confined to a well-defined path through the beam if they are going, one by one, to receive identical illumination as they flow through a round beam of light.

To alleviate this stringency, the beam-focusing lenses most frequently used in today's cytometers are a compromise; cylindrical lens combinations provide an elliptical spot, for example, 10–20 by 60 μm in size, with the short dimension parallel to the direction of cell flow and the longer dimension perpendicular to the flow (Fig. 3.1). By retaining a wide beam diameter across the direction of flow, an elliptical illumination spot can provide considerable side-to-side tolerance,

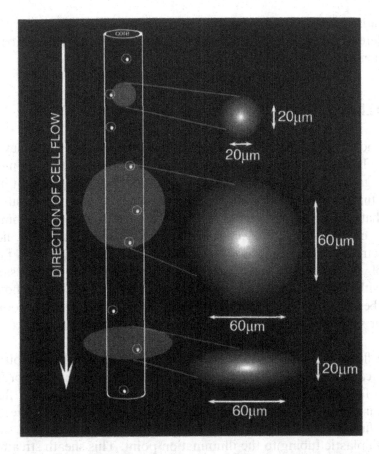

Fig. 3.1. Cells flowing past illuminating beams of different profiles. A beam with an elliptical profile (lower beam) allows cells to pass into and out of the beam quickly (avoiding the coincidence of two cells in the beam at the same time). In addition, it provides more equal illumination if cells stray from the center of the beam. The small circular beam at the top does not illuminate cells equally if they are at the edge of the stream core. The larger circular beam (in the middle) illuminates cells equally, but often includes multiple cells in the beam at the same time.

thus illuminating cells more-or-less identically even if they stray from the exact center of the beam; but at the same time this elliptical profile can provide temporal resolution between cells, illuminating only one at a time as they pass one by one into and out of the beam in its narrow dimension. The narrower the beam is, the more quickly will a cell pass through it—giving opportunity for the signal from that cell to drop off before the start of the signal from the next cell in line and avoiding the coincidence of two cells in the beam simultaneously. In multilaser systems, each beam of light is focused in a similar way but at different points along the stream; a cell moves through each beam in sequence.

CENTERING CELLS IN THE ILLUMINATING BEAM

The fluidics in a flow cytometer are likely to be ignored until they go wrong. If they go wrong disastrously, they can make a terrible mess. If they go wrong with subtlety, they may turn a good experiment into artifactual nonsense without anyone ever noticing. On the assumption that the more disastrous problems can be solved by a combination of plumbing and mopping (both essential skills for flow cytometrists), I will concentrate on the more subtle aspects of fluid control. Nevertheless, the potential hazard of working at the same time with volumes of water and with a high-voltage source should never be far from the mind of anyone working with a water-cooled laser or with a sorting cytometer with high-voltage stream deflection plates.

The flow on a flow cytometer begins (Fig. 3.2) at a reservoir of liquid, called the *sheath fluid*. Sheath fluid provides the supporting vehicle for directing cells through the laser beam. The sheath fluid reservoir is pressurized, usually with pumped room air, to drive the sheath fluid through a filter to remove extraneous particles and then through plastic tubing to the illumination point. This sheath stream is usually buffer of a composition that is appropriate to the types of particles being analyzed. For leukocytes or other mammalian cells, this usually means some sort of phosphate-buffered saline solution. Other cells or other particles may have other preferences.

Different instruments employ different strategies for getting the sample with suspended cells into the sheath stream in the cytometer. Some instruments require cells to be in small test tubes that form a tight seal around an O-ring on a manifold. The manifold delivers air to the test tube, thus pushing the suspended cells up out of the test tube and through a plastic line to the sheath stream. Other instruments use a motor-driven syringe to remove a volume of sample and then inject it slowly into the cytometer. Depending on the instrument, there may be a greater or lesser degree of operator control over the rate of flow. The amount of pressure driving the sample through the system will affect the uniformity of alignment between the cells and

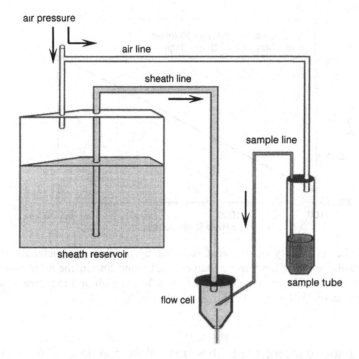

Fig. 3.2. The fluidics system, with air pressure pushing both the sample (with suspended cells) and the sheath fluid into the flow cell.

the illuminating laser beam as the cells move through the cytometer. Low pressure is less likely to cause perturbation of the stream profile and of the position of the cells within that stream. Empirically, if increasing the pressure on the sample causes undue broadening or wavering of signals, the pressure is probably excessive.

If cells flow too slowly through the cytometer, people start to make bad jokes about how microscopes cost less and are quicker. Because increasing the pressure may not be possible and, even if it is, is probably only a good idea within reasonable limits, the best way of getting cells to flow at reasonably fast rates is simply to make up the original sample with cells at a reasonably high concentration. A million cells per milliliter is often about right; 10^5 cells per milliliter is beginning to be low enough to test one's patience; 10^4 cells per milliliter is probably too low a concentration to be worth analyzing.

If you have few cells, make them up in a small volume (you will know how small a volume your system can handle). If the cells end

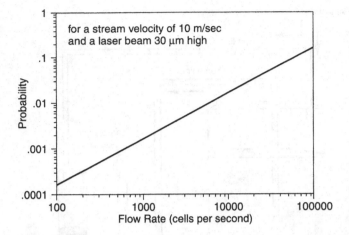

Fig. 3.3. The probability of an event recorded by the flow cytometer as a single "cell" actually resulting from more than one cell coinciding in the laser beam. For this model, the laser beam was considered to be 30 μm high and the stream flowing at 10 m per second.

up being too concentrated, they may flow too fast—but you can always dilute them on the spot and run the sample again. You may wonder why too rapid a flow is a source of problems. Faster seems as if it should always be better (especially around 5:00 PM). However, if cells are too close together as they flow through the laser beam, there may be difficulty separating their signals: A second cell may arrive in the illuminating beam before the preceding cell has emerged, and they will be measured together as if they were a single particle (with double the intensity). Figure 3.3 gives an indication of the probability of cells coinciding in the laser beam at different flow rates. Most cytometers seem to be quite happy looking at particles that are flowing at a rate of about 1000 particles per second.

Aside from concentration, another problem with samples is that the particles may be the wrong size. If they are too small, they may not be distinguishable from noise; nevertheless, bacteria and picoplankton and other bits and pieces of about 1 μm diameter or smaller are analyzable in at least some well-tuned cytometers. If the particles are too large, however, they will obstruct flow. If the fluid system is fully clogged, it may be difficult to get things flowing again; if it is only partially clogged, cells may flow but that critical alignment between stream and light beam may be skewed, thus causing arti-factual signals. Most experienced flow cytometrists recommend fil-tering any samples that are likely to contain large or clumped mate-rial before attempting to run them through the instrument. Nylon mesh of specified pore size works well (35 μm mesh is appropriate for most applications).

The exact size of particle that will be large enough to cause obstruc-tion depends primarily on the diameter of the orifice of the nozzle or flow cell being used (usually between 50 and 250 μm). This brings us to the next stop downstream in our following of the flow in this flow cytometer. The terms *nozzle, flow cell*, and *flow chamber* derive from different engineering designs for the best way of delivering cells into the sheath fluid and thence to the analysis point where they are illu-minated by the light. What we require is a method for keeping the cells in the center of the fluid stream so that they will pass through the center of the focused light beam and be uniformly illuminated. The flow chamber is the place in the cytometer where the cells from the sample join the fluid from the sheath reservoir. Within the flow chamber, the sample is injected into the center of the sheath stream; the combined sample and sheath streams are then accelerated as they move through a narrowing channel. This acceleration is critical to the precise alignment of cells, one at a time, in the laser beam. References at the end of this chapter will direct interested readers to mathemati-cal discussion of the fluid mechanics related to this subject. For our purposes here, it will be enough to note that, as a result of consid-erations pertaining to laminar flow through a narrowing path, the sample with its suspended cells, after injection into the center of the sheath stream, will remain in a central core as it flows within the sheath stream out through the flow chamber.

The technical term for this is *hydrodynamic focusing*; flow of a sample stream within the center core of a sheath stream is called *coaxial flow*. The exact diameter of that central sample core within the sheath stream is related to, among other things, the rate at which the sample is injected into the sheath stream; a 100 μm sheath stream may, depending on sample injection velocity, have a core width of perhaps 5–20 μm (Fig. 3.4). Because hydrodynamic focusing tends to confine the cell sample to this central core, there is little mixing of sample with sheath fluid (but diffusion of small molecules will occur). The reason that this type of coaxial sample flow suits flow cytometry

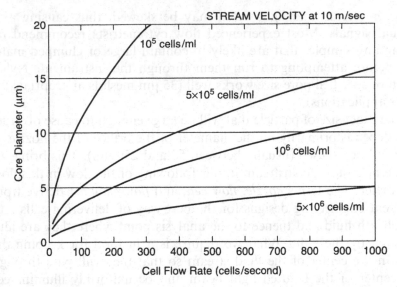

Fig. 3.4. When cells are pushed through the cytometer at faster rates, the sample core widens. Use of a more concentrated cell suspension allows a faster flow rate while maintaining a narrow core.

is that a nozzle or flow cell with a relatively wide orifice (50–250 μm) can be used to avoid blockage; the particles to be analyzed are nevertheless maintained in alignment in the narrow (say 5–20 μm) core so that they progress in single file down the center of the stream through the laser beam. This ensures uniform illumination as long as the central core is narrow and the illuminating beam is focused at the center of the sheath stream. It also ensures that, because the cells are stretched out at a distance from each other as rare beads along a gold chain, for the most part one particle is illuminated at a time.

We are now in a position to understand why radical changes in the rate at which the sample is pushed through the flow cell will cause changes in the resulting light signals. Changes in sample injection rate cause changes in the diameter of the core; when the size of the core increases, particles are no longer so tightly restricted in their position as they flow past the light beam, and illumination will be less uniform (Fig. 3.5). The size of the core often has critical impact in this way on DNA applications where precise analysis is important and nonuniform illumination causes nonuniform fluorescence. The stream diameter may be 100 μm, but if the illuminating beam has a width of 60 μm, then the core diameter needs to be considerably less than

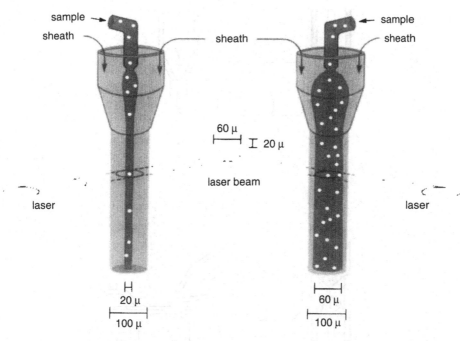

Fig. 3.5. The flow of cells within the core of sheath fluid through the analysis point in the illuminating beam. When the sample is injected slowly (left), the core is narrow and the cells flow one at a time through the center of the laser beam. When the sample is injected too rapidly (right), the core is wide (somewhat exaggerated in this drawing); the cells may be illuminated erratically because they can stray from the center of the beam. In addition, more than one cell may be illuminated at the same time.

60 μm to keep the cells at the center of the beam in order to ensure uniform illumination. In addition, with a wide sample core, two cells are more apt to be illuminated simultaneously, resulting in addition of their signals.

In some systems, the sheath stream with central sample core emerges from the orifice of a nozzle where it is then intersected by the light beam in the open air (a "jet-in-air" configuration). In other systems, the stream is directed (either upward or downward) through a narrow, optically clear, flow cell or chamber; the particles are illuminated by the light beam while they are still within this flow chamber. In still other systems, a nozzle forces the stream at an angle across a glass coverslip (Fig. 3.6). All systems have their advocates—the positive and negative considerations are based mostly on ideas about signal

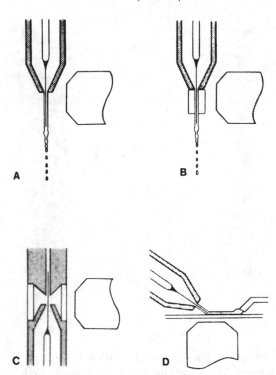

Fig. 3.6. Various types of flow chambers. **A** and **B** are designs used in sorting cytometers (in A the analysis point is in air after the stream has left the flow cell; in B analysis occurs within an optically clear region of the chamber itself). C and D are two designs for nonsorting cytometers (in C the stream flows upward through an optically clear region of the chamber; in D the stream is directed at an angle across a glass coverslip). Adapted from Pinkel and Stovel (1985).

noise, stream turbulence, and the control of drop formation for sorting. There is no clear favorite, and purchasers of commercial instruments usually base their choice of cytometer on factors other than flow cell design and then live with the design they get.

The main point of concern on a daily basis is the avoidance of blockage. Most flow cells in nonsorting cytometers have orifices with relatively wide dimensions (perhaps 150–250 µm in diameter) and are tolerant of large material; sorting instruments are more restricted in nozzle size. Although instruments can be modified to sort large cells, most commercial sorting cytometers operate with a stream diameter of between 50 and 100 µm. Cells larger than the nozzle orifice diameter will clog the nozzle. Furthermore, even if you think you have a suspension of single cells, there will undoubtedly be some aggregates of much larger size.

DETECTION OF SIGNALS FROM CELLS

An optical bench is simply a table that does not wobble. A flow cytometer's optical bench may be visible at the back or may be incorporated behind a closed door; in either case, it provides a stable surface that fixes the light source and the light detectors in rigid alignment with the objects being illuminated. If the bench is moved, the light source, the light detectors, and the object of illumination will move in synchrony so that alignment between the three does not change. The reason that users of a flow cytometer should know about optical benches is, simply, to remind them that signals from a cell can vary beyond recognition if this alignment changes even slightly.

Figure 3.7 is a diagram of the components that sit on the optical

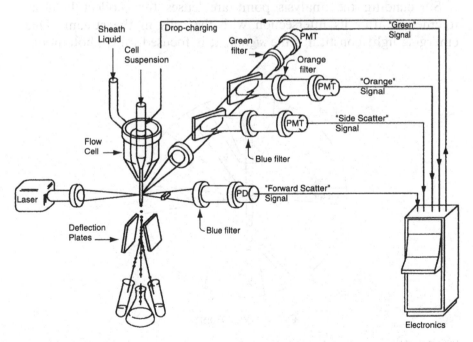

Fig 3.7. Components on the optical bench of a generalized "four-parameter" flow cytometer. (The drop charging, the deflection plates, and the drops moving into separate test tubes apply only to sorting cytometers [see Chapter 9] and not to benchtop instruments.) Adapted from Becton Dickinson Immunocytometry Systems.

bench of a flow cytometer. If we follow the light path from the beginning, we can see that the light, after it leaves the laser source, is
focused through a lens into an elliptical beam of about 20 by 60 μm
as it approaches the liquid stream. The stream of 50–150 μm diameter flows perpendicularly to the light beam. The alignment of the light
beam and the stream must ensure that the stream is intersected by the
light beam in such a way that the core of the stream (with its cells) is
uniformly illuminated by the light. The point at which stream and
light beam intersect (Fig. 3.8) is called the *analysis point, observation
point*, or *interrogation point*. If the light beam and the stream are not
perfectly and squarely aligned with each other, then cells within the
stream will be erratically illuminated and will give off erratic light
signals. Although the reasons for imperfect alignment may have to do
with poor adjustment of the focusing lens or light source, the stability
of the optical bench is, on the whole, reliable. On a day-to-day basis,
poor alignment is much more likely to result from shifts in the fluid
stream resulting from bubbles or partial obstruction.

Surrounding the analysis point are lenses that collect light as
it emerges after its intersection with the cells in the stream. This
emerging light constitutes the *signal*. It is focused onto photodiodes

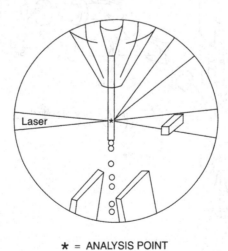

★ = ANALYSIS POINT

Fig. 3.8. The analysis point. Alignment between the illuminating beam and fluid
stream is critical in determining the characteristics of the resulting forward- and
right-angle signals.

or photomultiplier tubes (the latter are more sensitive to dim light than the former; both can be called *photodetectors*) that convert the light signal into an electrical impulse. The intensity of the electrical impulse is proportional to the intensity of the light impinging on the detector.

The photodiode that sits in direct firing line of the illuminating beam receives signals that are responsible for much misunderstanding. Rather than simply detecting the amount of the illuminating beam (usually turquoise blue color) that gets through after bombarding a cell in the stream, this forward-angle photodetector is fitted with a so-called obscuration bar in front of it. The function of this narrow metal strip is to block the illuminating beam so that it cannot reach the detector. It may be wondered why we bother to place a photodetector in this position on the optical bench if we are then going to prevent any light from hitting it by blocking that light with a strip of metal. The critical point is that any light that has been bent as it passes through or around a cell will manage to avoid the bar, strike the photodiode, and generate a signal. Depending on the width of the bar, the angle of bending required to generate a signal is usually about $0.5°$. This so-called forward scatter (FSC) or forward-angle light scatter (FALS) signal (defined as light of the same color as the illuminating beam that has been bent to a small angle from the direction of that original beam) is sometimes called a *size* or *volume* signal. It is undoubtedly related to the size and volume (mostly to the cross-sectional area) of the cell, but it is also related to other factors such as the refractive index of the cell. A cell with a refractive index very different from that of the surrounding medium will bend more light around the bar than will a cell with a refractive index closer to that of the stream. It is true that a large cell will bend more light than a small cell of the same refractive index; however, confusion inevitably results when the term *volume* is used carelessly to describe this forward-angle scatter.

For example, anyone used to looking at cells under a microscope will say that dead cells appear larger than living cells of the same type. It is therefore something of a surprise to find that dead cells appear "smaller" than living cells in a flow cytometer. Of course, they are not actually smaller; they simply give a dimmer forward scatter signal than living cells because they have leaky outer membranes, and the refractive index of their contents has for this reason become more like the refractive index of the surrounding stream. They therefore bend less light into the FSC detector than a viable cell does. To give one more example (just to hammer the message home), erythrocytes, despite their uniform volume, give a broad range of forward scatter signals in a flow cytometer because their cross-sectional area varies depending on their orientation in the stream. However, anyone who remembers Fulwyler's experiments with red cells in a Coulter counter should not find this fact too surprising.

Since I mention Coulter volume measurements, I should add, for the sake of completeness, that some flow cytometers (most notably the now-obsolete Becton Dickinson FACS Analyzers) have a volume signal that is based on the Coulter-type measurement of electrical impedance. This is not in any way related to the FSC signal discussed in the paragraphs above. The Coulter-type volume signal is proportional to the volume of a particle (as well as to its electrical characteristics). The FSC signal is proportional to the cross-sectional area of a particle (as well as to its refractive index). Some state-of-the-art cytometers actually record both kinds of signals from particles; the ratio between the two varies for different types of particles, and this can be instructive. The take-home message is, therefore, that both of these kinds of signals give information about the physical characteristics of a particle, but neither tells much that we can count on about the true volume of that particle. The news, however, is not all bad. Although we may not know the exact biological meaning of a forward scatter signal, scatter characteristics can allow us to distinguish different classes of cells from each other, and this can, as we shall see, be useful.

In addition to this detector in the line of the laser beam (to detect FSC light), the standard configuration for photodetectors around the analysis point in a flow cytometer includes three, four, or more photomultiplier tubes at right angles to the illuminating beam Because they are at the side and not in direct line of the illuminating beam, no light will hit these detectors unless something in the stream causes the illuminating beam to be deflected to the side or unless something in the stream is in itself a source of light in that direction. One of these detectors is usually fitted with a glass filter of the same color as the illuminating light (say, turquoise blue); it will register any illuminating (turquoise) light that is bounced to the side from the surface of a

cell in the stream. The rougher or more irregular or granular a particle is, the more it will scatter the illuminating beam to the side. The
intensity of this so-called side scatter (SSC) light (defined as light of
the same color as the illuminating beam that is scattered by a particle
to an angle of 90° from the illuminating beam) is therefore related to
a cell's surface texture and internal structure as well as to its size
and shape. It is sometimes referred to as a *granularity signal* or an
orthogonal light scatter signal. Granulocytic blood cells with their granules and irregular nuclei, for example, produce much more intense
SSC than do the more regular lymphocytes or erythrocytes.

The several other photomultiplier tubes that sit at right angles to
the illuminating beam are there to detect other colors of light (for
example, green, orange, or red) that might be emitted by the cell. If
the illuminating beam is blue, then there is no way that green, orange,
or red light will emerge from the analysis point unless a cell in the
stream is itself generating that green, orange, or red light. A cell can
generate light either because it contains endogenous fluorescent compounds or because a scientist has stained it with a fluorescent stain. A
cell stained with a fluorescent stain will, when illuminated from one
direction by a beam of blue light, emit in all directions light of a different color (the color depending on the fluorescent stain used). It is
this fluorescence that will be registered on one or another of the right-
angle photomultiplier tubes.

The light coming out at right angles from the analysis point will be
a mixture of light of the same color as the laser beam (SSC) and light
of an assortment of other colors, depending on the fluorescent stains
or endogenous molecules associated with the cell. Photodetectors are,
unfortunately, relatively color blind; within their working range, they
register all wavelengths of light more or less equally and give out
electrical impulses that are proportional to intensity but provide no
information about the color of the detected light. It is by restricting
each photodetector, in some way, so that it receives only a particular
color of light that we can associate the signal from that detector with
a color and thereby obtain any specific knowledge about the color of
our cell.

Therefore, in order for the signal from a particular photodetector
to tell us what color a cell is, we need to use colored filters to make
the photodetectors color specific. Photodetectors can be fitted with,

for example, green, orange, or red filters—depending on how many of these detectors are present in a particular instrument. The role of the filter is to ensure that each photodetector "sees" only light of the color transmitted through its own filter. If a green filter is in front of photodetector number one and an orange filter is in front of photodetector number two, then pd1 will respond with an electrical impulse if green light emerges from a cell at the analysis point, pd2 will respond with an electrical impulse if orange light emerges from that cell, both detectors will respond if white light emerges, and neither will respond if blue, or yellow, or no light appears. In this way, the signal from each photomultiplier tube can indicate the presence of a particular fluorochrome on a particle.

By way of recapitulation, before we move on to electronics, recall simply that, in the flow chamber, the sample with its suspended cells has joined the sheath fluid, and, within the core of that sheath stream, the cells have (with perfect orthogonal alignment) reached the intersection with the illuminating light beam. When the focused light beam hits the cell in the stream of fluid, the light is affected in several ways. It can be reflected, diffracted, and/or refracted; it can also be converted to a different color if it has been absorbed by a fluorescent chemical. The light that emerges after hitting a cell may register on one or more of the available photodetectors. The photodetectors each measure some specific aspect of the emerging light because of their positions, the color of their filters, or the presence of an obscuration bar. In a typical configuration, two of the photodetectors measure forward-angle scatter light and side scatter light in order to provide some information about the physical characteristics of the cell (sometimes called *size* and *granularity*); and three or more photodetectors are equipped with colored filters to provide information about the fluorescent light being emitted by the cells. These five or more characteristics registered on the photodetectors as each cell passes through the light beam are known as the *measured parameters* in the flow cytometry system. An instrument may, for example, be called a *three-parameter cytometer* or an *eleven-parameter cytometer* depending on how many photodetectors are arranged around the analysis point. Most commercial instruments now have a five- or six-parameter configuration. Once the light has hit the photodetectors, the light signals are converted into electrical signals.

ELECTRONICS

For many reasons, the circuit boards of a flow cytometer are things of beauty with which we are not going to concern ourselves. Integrated circuits (chips) do occasionally wear out; if they do, the symptoms can be most peculiar and difficult, even for a trained cytometer engineer, to diagnose (in our lab, we once had side scatter signals masquerading as green fluorescence). The message here is that you should get to know the engineer who services your instrument, and you should be nice to her at every possible opportunity. You will almost certainly need her expertise some day. In this section, however, I will describe just enough of the electronics so that you can understand the control that you can exert over the way the light signal given off by a particle reaches the computational circuitry for data analysis. Your ability to use these electronic controls correctly will have a significant influence on the interpretability of your data.

Photodetectors, as I have said, convert light signals into electrical impulses. The intensity of those electrical impulses are, within the limits of normal operation, related to the intensity of the light signals. However, the way in which the electrical impulses are then treated so that they are strong enough to be processed is open to user choice. Photomultiplier tubes (but not photodiodes) have voltages applied to them so that the cascade of electrons resulting from the original light impulse is converted into a sufficiently large current to be measured. Changing the voltage applied to a photomultiplier tube is one way we have of increasing or decreasing the current response of that tube (if we go too high or too low on the voltage, the response of the tube will no longer be proportional to the amount of light received). The second method we have at our disposal to increase or decrease the response to light is to change the amplification of that electrical current after it leaves the photodetector. Amplifiers can be either logarithmic or linear and can operate at varying gains. In general, we should be aware of the fact that a logarithmic amplifier allows us to look at light signals over a wide range of intensity; a linear amplifier may restrict our sensitive measurements to signals all in the same range.

Thus our two types of controls on the photodetectors are, one, the control of the voltage applied to photomultiplier tubes and, two, our

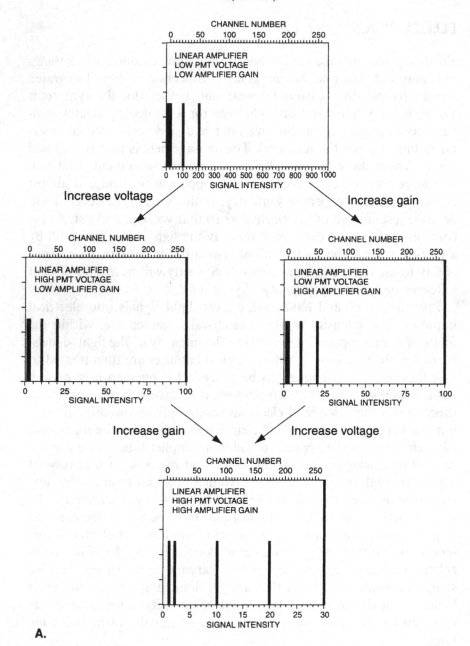

A.

Fig. 3.9. The effect of changes in the photomultiplier tube voltage and amplifier gain on the appearance of six signals with intensities in the relationship of 1:2:10:20:100:200 to each other. **A:** Linear amplification. **B:** Logarithmic amplification. (*continued on next page*)

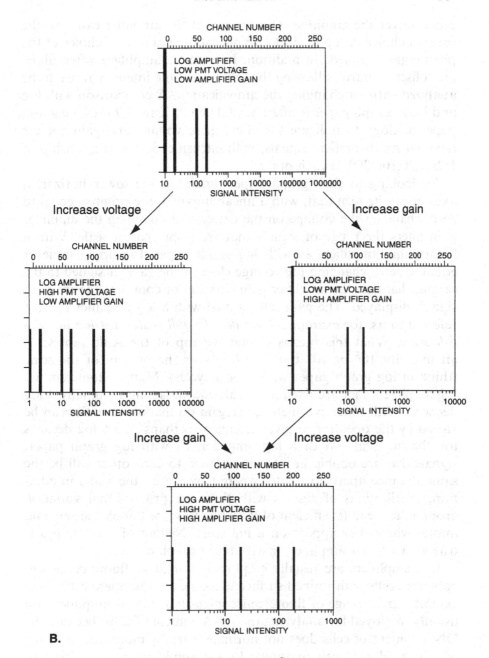

B.

Fig. 3.9. (*continued*)

control over the amplifier. Our control of the amplifier can take the form of choice of log or linear amplification and also of choice of the precise gain applied. In addition, logarithmic amplifiers often allow an "offset" control, allowing the selection of the intensity range to be analyzed without changing the amplification. A comparison with log and linear graph paper is often helpful here. Figure 3.9 uses the graph paper analogy to indicate the effect these voltage and gain choices have on six theoretical signals, with intensities in the relationship of 1:2:10:20:100:200 to each other.

By looking at the "graph paper" scale on the lower horizontal axes, it can be seen that, with a linear amplifier, the origin is equal to zero; changing the voltage on the detector or changing the amplifier gain alters the range of signals that are displayed on scale. With a logarithmic amplifier, as with log graph paper, the origin is never equal to zero; changing the voltage changes the value assigned to the origin; changing the amplifier gain stretches or contracts the range of signals displayed. The gain setting used with a log amplifier is often referred to as, for example, *3 log decades full scale* or *4 log decades full scale*. What this means is that the top of the scale represents an intensity 10^3 or 10^4 times as bright as the bottom of the scale (think of log graph paper with 3 or 4 cycles). Many cytometers use log amplifiers that are fixed at a scale encompassing nominally 4 log decades. With other cytometers, the gain on the log amplifier can be varied by the operator to give a range of perhaps 2 to 5 log decades for the full scale. With a log amplifier, as with log graph paper, signals that are double in intensity relative to each other will be the same distance apart no matter where they are on the scale; in addition, distributions of signals with the same proportional variation around the mean (coefficient of variation [CV]) will look the same no matter where they appear on a log scale. Neither of these things is true with a linear amplifier or with linear graph paper.

Log amplifiers are usually employed to analyze fluorescence signals from cells with stained surface markers, because these cells often exhibit a great range of fluorescence intensities. Linear amplifiers are usually employed for analyzing the DNA content of cells, because the DNA content of cells does not normally vary by more than a factor of 2 (e.g., during cell division). Linear amplifiers may be used to analyze forward and side scatter signals, but practice here is apt to vary from lab to lab. With either linear or logarithmic amplification, the choice of voltage or amplifier gain will depend on the range of intensities to be analyzed. The goal is to choose electronic settings so that the dimmest cells analyzed fall at the low end of the scale and the brightest cells fall at the top end of the scale.

Once the electrical signal from a photodetector has been amplified, its intensity is then analyzed and this value will be recorded by means of an analog-to-digital converter (an ADC). The role of an ADC is to look at a continuous distribution of signals and group (or bin) them into discrete ranges—much as you would need to do if you wanted to plot the heights of a group of people on a bar graph (histogram). For a height distribution histogram, the first bar might represent the number of people with heights (in old-fashioned feet and inches) between 4'9.5" and 4'10.5" (4'10" ± 0.5"); the second bar, the number of people with heights between 4'10.5" and 4'11.5"; and so on (Fig. 3.10). Similarly, the ADC of a flow cytometer divides the electrical signals that it receives from each photodetector into discrete ranges.

Here we run into that term *channel* that flow cytometrists use so often. The ADCs on a flow cytometer are divided up into a discrete number of channels; the number of channels is usually 1024 (but may be 256 or even 65,536). Each channel represents a certain specific light intensity range (like the 17 channels each with a 1 inch range on our cadet height example). The signal from a cell is recorded in one or another channel depending on the intensity of that signal. It is the combination of photodetector voltage and amplifier gain that allows the user to set the intensity range represented by each of those channels. Using, for our example, a 256 channel ADC, we can now look back at the upper horizontal axis in each histogram in Figure 3.9. Having used the graph paper scale on the lower axis to plot intensity, we have now divided up the whole range of light intensity into 256 channels on the upper axis (numbered from 0 to 255). Varying the voltage and gain simply assigns a different intensity to each of the channels. The goal is to have the full range of channels encompass the full range of intensities relevant to a particular experiment. When the voltages and gains have been selected for a given experiment, the output data are then simply recorded by the cytometer electronics as light intensity on a scale of 0 to 255 (or 1023).

It is only by knowing something about the amplifier and voltage settings that you can know what relationship those values of 0 to 255 or 1023 bear to one another. For example, if we are using an ADC

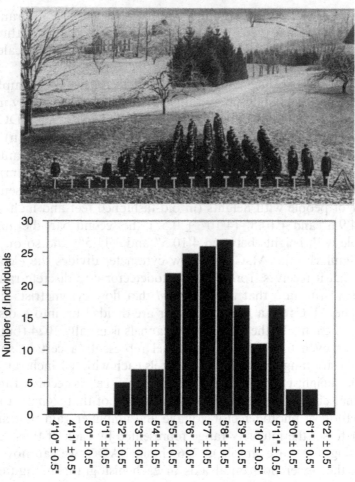

Fig. 3.10. World War One cadets from Connecticut Agricultural College arranged to form a height histogram. Photograph from A. Blakeslee (1914).

with 1024 channels and have selected a log amplifier with a gain that gives us 2 decades for the full scale, then we would know that a signal appearing in channel 700 has been given off by a cell that is 10 times as bright as a cell giving a signal in channel 188 (with a 2 decade full scale amplifier, every 512 channels represents a 10-fold increase in intensity; the entire 1024 channels represent two consecutive 10-fold increases in intensity $= 10^2$). If, on the other hand, the gain on the log amplifier had been set to give us 4 log decades for the full scale, then a signal in channel 700 would come from a cell that is 100 times brighter than a cell giving a signal in channel 188 (on a 4 decade scale, every 256 channels represent a 10-fold increase in intensity; 512 channels represent a 10^2-fold increase; the entire 1024 channels represent a 10^4-fold increase). If we had been using a linear amplifier, then a signal in channel 700 would represent a cell 3.72 times brighter than one with a signal in channel 188 ($700/188 = 3.72$).

These examples should serve to emphasize that the numerical "read out" from a flow cytometer is relative and user-adjustable. A knowledge of instrument electronics and the settings used is required if we really want quantitative information about the relative brightness of signals from different cells; a simple channel number is not really enough. As an example, Figure 3.11 indicates real data (compare with Fig. 3.9 of model data) acquired with a mixed suspension of the same set of particles (beads of five different intensities) but with two different electronic settings (a linear amplifier in the histogram above and a log amplifier below). The channel numbers for the five peaks and the distribution of the peaks across the 1024 channels are very different in the two cases (next time you run beads or cells on a cytometer, try this yourself).

One other capability we have with cytometry electronics is the definition of a *threshold*. An electronic threshold is just like a threshold into a room: It defines an obstacle. Only cells giving signals greater than that obstacle will be registered on the cytometer ADC. The most common use of this threshold is in the definition of a forward scatter channel number. Only cells with a forward scatter signal brighter than the defined channel threshold will be registered by the cytometer. With the use of a forward scatter threshold, we can avoid problems that might come from dust, debris, and electronic noise in the system. The dim forward scatter signals from debris and noise are not bright enough to pass over the threshold, and "particles" of this type would be completely ignored. There are other ways of using a threshold (for instance, by using a red fluorescence threshold to exclude from an ocean water sample the signals from any particles that do not contain chlorophyll), but the use of a forward scatter threshold is by far the most common. Thresholds should be used with care; setting the threshold level too high can hide important data and can even make you think that you have no cells in your sample.

Flow Cytometry

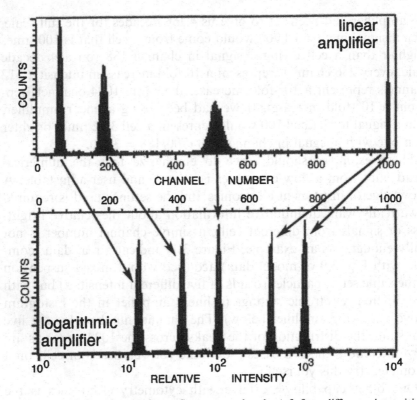

Fig. 3.11. Intensity signals from fluorescent beads (of five different intensities) acquired with linear amplification (top) and logarithmic amplification (bottom). Log amplification permits all five intensities to be "on scale" (that is, within the 1024 channel range). Additionally, the spread (the CV) and the peak height of the distributions for each bead are visually similar with a log but not with a linear amplifier. From Givan (2001).

By now, we should have a reasonably clear picture of the physical and electronic characteristics that form the basis for flow cytometric analysis. We have followed cells into the center of a stream flowing through a nozzle into a light path; we have seen how reflection, refraction, and fluorescence can generate light signals from those cells as they are hit by that light beam; we have accounted for registering of the light emerging from the cells onto one or another photodetector depending on the color or direction of that emerging light and the filters in front of the individual photodetectors; and we have described the way that the intensities of the light from cells registered on each photodetector can be assigned to the discrete channels (1024) of an ADC. Therefore, for each cell that has flowed past the illuminating beam, we now have, simply, four or five or more numbers (depending on the number of photodetectors present) that describe that cell. Those four or five numbers (each on a scale of 0 to 1023) tell us the intensity of the FSC, SSC, and fluorescence (red, green, orange, and so forth) from that cell. Those numbers are, quite simply, the only information we now have about that cell. These are the facts that can be correlated with each other for data analysis. What we now do with the numerical data is up to the computer software and hardware available.

FURTHER READING

Chapter 2 in **Melamed et al.** and Chapters 1 and 4 in **Shapiro** are good general descriptions of cytometer characteristics.

Chapter 3 in **Melamed et al.** and Chapter 3 in **Van Dilla et al.** discuss hydrodynamics and flow chamber design in depth.

Chapter 2 in **Watson** has a good discussion of fluid flow dynamics and of the avoidance of coincidence events.

Chapter 5 in **Darzynkiewicz** covers flow cytometric optical measurements in general.

4

Information:
Harnessing the Data

DATA STORAGE

Having left our flow cytometry system with light signals from a single cell recorded in appropriate channels on the analog-to-digital converter (ADC), we are now faced with the prospect of losing all these data as soon as we start recording data from our next cell. What is required is a way to store the data permanently for later correlation and analysis at our leisure. At this point we leave cytometry *sensu stricto* and find ourselves in the realm of computer ware (soft and hard).

The main challenge encountered with storage of data from flow cytometry arises from the ability of a flow system to generate large amounts of data very quickly. In a four-parameter configuration (four photodetectors), each particle flowing past the light beam generates four signals. If we want to analyze 10,000 cells from each sample (this sounds like a lot to someone used to microscopy, but flow cytometrists can analyze 10,000 cells in, say, 10 seconds, and therefore are easily persuaded that a large number of cells gives statistically better information), then each sample that is run through the flow cytometer will generate 40,000 numbers. If each of those numbers covers the range of 0–1023, then 10 bits ($2^{10} = 1024$) of information are required to specify each of those numbers. Because bits come in packets of 8 (and 8 bits are known as one "byte"), we need two bytes of storage space to specify the intensity of each parameter. This comes to 80,000 bytes for a file that describes four parameters about each of 10,000 cells (plus a few extra bytes for housekeeping arrangements). A six-parameter cytometer will generate proportionally more data, that is, 120,000 bytes from that same sample.

Therefore (staying with our downmarket four-parameter data file), the information from just seventeen 10,000-cell samples will fill a 1.4 MB floppy disk. Floppy disks are readily available and may be the answer to data storage problems if the experiments are small; they will be an expensive and cumbersome answer if the experiments are large. Although all flow cytometrists start out with small experiments, most progress rapidly to experiments with 20 or 30 or more samples (think of using 96 well microtiter plates for processing cells). In fact, one of the surest rules of flow cytometry (a rule even better known to imaging scientists) is that data will, in less time than predicted, fill all available storage capacity. When you find yourself continually running out of floppy disks, you will begin to try to think of other options for data storage. The other options for flow cytometry data storage are just the same as for any other kind of computer data storage: These options change with time, but currently include, for example, hard drives, zip cartridges, and CD-ROMs (Table 4.1). Options at any given computer will be determined or limited by the peripheral hardware available, but most systems will have a large-

TABLE 4.1. The Number of Flow Data Files (10,000 Cells/4-Parameter Data/1024-Channel Resolution) that Can Be Stored to Various Types of Media[a]

Medium	Capacity	Cost (US$)	Cost per MB (US$)	Number of files per disk
Floppy disks	1.4 MB	$0.50	$0.36	17
Zip cartridges	100 MB	$10	$0.10	1,250
	250 MB	$15	$0.06	3,125
Jaz cartridges	1 GB	$100	$0.10	12,500
Hard drives	10 GB	$300	$0.03	125,000
	50 GB	$1200	$0.02	625,000
CD-ROMs	600 MB	$1	$0.0017	7,500

[a]The "number of files per disk" is given as the number of samples (of 10,000 cells each) whose list mode data (4-parameter/1024-channel resolution) after acquisition can be stored to media of the indicated size. The capacity in bytes of the different media is representative but will vary with the formats of different computing systems. Similarly, different acquisition software will require more or less extra storage space for the housekeeping information that is stored with each sample. Prices of media are illustrative, but will vary considerably from place to place and over time.

capacity hard drive, which might store the information from many samples. Although increasingly inexpensive, hard drives have two main drawbacks. The first is that you cannot take them home with you, and therefore someone else who has access to the system can trash your data (and, given enough time, probably will do just that). The second problem with a hard drive is derived from that rule about data filling all available storage capacity: No matter how large the capacity of the hard drive, it will become full sooner than expected.

The solution to both of these problems is to have removable back-up capability. This back-up device could consist of zip cartridges or floppy disks; both of these options are appropriate for immediate data storage. For long-term archiving, the least expensive (and slowest) option is computer tape. Recordable CD-ROMs currently combine low cost, moderate speed, and a reputation for stability; the price of rewritable CD-ROM burners has come down, but generally this medium has been used for permanent archiving on write-once disks. For example, when six zip cartridges have been filled, their data could be transferred permanently to a CD-ROM and the zip cartridges re-used. Considerations in choice of medium will involve the cost of the medium, the cost of the drive, the speed of writing and accessing the data, and, importantly, the convenience of organizing data files on disks of various sizes. With any luck, you will have backed your data from the hard drive to the removable medium of choice and will have the data safely in your pocket when someone "accidentally" clears the system. For these reasons, back-up capability of some type is especially necessary for multiuser flow cytometers.

DATA ANALYSIS

Now that data have been stored, we come to analysis, which is the real point of everything we have done so far. Methods for data analysis vary. They vary with the inclinations of the software programmer; they also vary with the budget of the cytometer facility. They may be strictly commercial, or they may be homemade. These days, commercial manufacturers of cytometers compete with each other on the basis of their software systems as much as on the basis of their cytometer technology. In addition, independent entrepreneurs, with increasing frequency, have begun to program for analysis of flow data and have successfully entered into the commercial market. Moreover, there are people (for example, various scientists at the Los Alamos Labs in New Mexico or Joe Trotter at the Scripps Research Institute, San Diego, California) who have developed flow analysis software that they distribute without charge. It is increasingly true that the software available for analysis plays a large role in what the user sees as an acceptable cytometer package. Samples may be run through a cytometer and the information from those particles stored very quickly; analysis and re-analysis of that information may then require a great deal of time. Therefore, it is not surprising that software is an important aspect of flow cytometry.

In theory, data from all flow cytometry systems are now stored in so-called flow cytometry standard (FCS) format. This means that, although data stored after acquisition on one cytometer may not be analyzable on software from another cytometer (because manufacturers have been discouragingly slow at fully embracing the standard), the format is one that can be learned, and, in principle, anyone with good programming skills could write software for analysis of flow cytometry data. The FCS format also means that independent programmers can and do write programs that will handle data acquired on any cytometer. In practice, most people use commercial packages for data acquisition and storage because these packages are commonly purchased along with the cytometer. Although the same packages provide methods for data analysis, there are times when additional analysis software from an independent source may be helpful. This might be to provide advanced analysis methods, to provide especially pretty pictures for slides and publication, to store data to a database, or to allow analysis on one or another home computing system.

The data stored in FCS format are usually "list mode" data. As described above, this means that, in a four-parameter cytometer, four numbers are stored for each cell. A 10,000 cell data file will consist of a long list of 40,000 numbers, with each set of four numbers describing each cell in the sequence in which it passed through the laser beam. By retrieving the stored data, each cell can be analyzed again, and the intensity of each of the four signals for that cell will be known and can be correlated with each other or with the intensity of the four signals from any (or all) other cell(s). This type of list mode data is useful because no cytometric information has been lost, and it can all be examined again in future computer analyses.

There are other types of data storage that have the advantage of requiring less storage space. So-called single-parameter data storage involves the storage of the intensity profiles for the population of cells in a sample for each parameter separately; the only information stored is, for example, the distribution of forward scatter signal intensities for the cells in the sample; the distribution of side scatter signal intensities for the cells in the sample; the distribution of red fluorescence signal intensities for the cells in the sample; and the distribution of green fluorescence signal intensities for the cells in the sample. In this case, however, no information has been stored about whether the bright green cells are the cells that are bright red or whether they are the cells that are not red. With this kind of storage, we will not know whether the cells with a bright forward scatter signal are red or green or both red and green. Therefore, if we have stored data as single-parameter information, we tie up little storage capacity but we severely restrict our options for future analysis. Unless storage capacity is very limited and the information required from data analysis is also very limited, list mode data storage is by far the best and indeed the only recommended option. Another rule about flow cytometry data analysis is that you are always going to want more information out of a sample than you thought was required when you planned and carried out the experiment. So use list mode data storage unless there is a very good reason for doing otherwise.

Having stored list mode data for all the particles in a sample, software allows the correlation of the data in all possible directions. We can readily look at any one parameter in isolation and analyze the light intensity histogram from the cells in a sample with respect to that parameter. Because we have stored the channel numbers characterizing the signals from each cell, the software can plot a histogram (number of cells at each channel as in the height histogram; see Fig. 3.10) with all the cells placed according to the channel number characterizing the intensity of their signals. In this way we could look at the intensity distribution of, for example, green fluorescence signals from 10,000 cells in one sample. And then we could look at the intensity distribution of red fluorescence signals from the same 10,000 cells. We can, in fact, generate a histogram for each of the parameters measured. Some software will plot data according to channel number; other software will convert the channel data and use a "relative intensity" scale (think of the upper and lower horizontal axes in Fig. 3.9). In the latter case, the software is making assumptions about the accuracy, linearity, and amplification gain on the photodetectors. Once we have plotted the histogram distribution (number of cells on the y-axis at each defined light signal intensity on the x-axis), the software will allow us to analyze this distribution to extract certain kinds of information: for example, what percentage of the cells fall within a specified intensity range, what the most common intensity (channel number) is for the group of cells (the "mode" channel), what the mean intensity channel is for the group of cells, or what the median intensity is for that group (Fig. 4.1).

Just how these values are obtained will vary with the particular software in question. "Cursors" or "markers" can be used to define regions of intensity that may be of interest. For example, we could place a cursor so that it separates the low-intensity range of green fluorescence from the high-intensity range of green fluorescence, and we could then ask how many cells fall within the high-intensity range. A value for the percentage of positively stained cells can be determined by placing a cursor at a position defined by the background fluorescence of unstained cells. By convention at the 1–3% level (that is, at an intensity that clips the bright edge of the unstained cells and makes 1–3% of them "positive"), this kind of cursor is usually the best way to describe a mixed population that consists of both unstained cells and brightly stained cells.

If we are, on the other hand, concerned with the changing fluorescence intensity of a uniform population of cells, we would be better served by using mode or median or mean characteristics of that population (the use of a "percent positive" value is, in fact, highly misleading if we are looking at a population of cells that are uniformly but dimly fluorescent). The mode value, being simply the channel number describing the intensity of the most frequent group of cells (the peak of the histogram), may vary erratically and be poorly reproducible if the population distribution is very broad. The mean value will be incorrect if significant numbers of cells are in the highest channel (255 or 1023) or lowest channel (0) of the histogram. The median value for the population is most reproducible because it

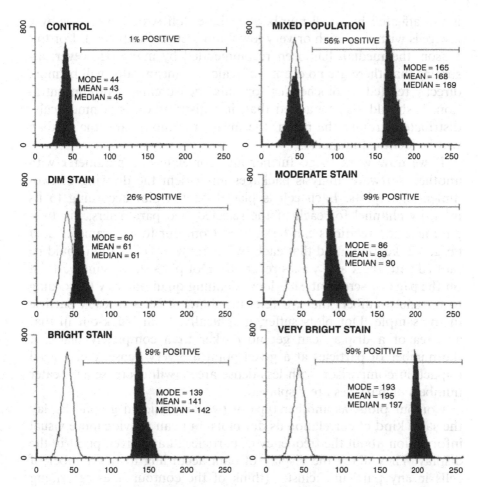

Fig. 4.1. Methods for describing the histogram distribution of signal intensities from a population of cells. The plots show the number of cells on the vertical axis against channel numbers (related to scatter or fluorescence intensity) on the horizontal axis. Control (unstained) cells are indicated as the clear distributions overlayed with the black distributions from the stained cells. If cursors or markers are placed to delineate a region of positive intensity (relative to the 1% level on an unstained control), the "% positive" value can usefully describe a mixed population of stained and unstained particles. This value will be misleading if used to describe a uniform population of dimly stained cells. The "mode," "mean," or "median" channel number can be used to compare uniform populations of cells of varying fluorescence intensity.

is not affected by small numbers of these "off scale" events or by a few cells with very high or very low fluorescence ("outliers"). For this reason, the median has been recommended by many. However, assuming that there are no events off scale, the mean value will be more directly related to biochemical or bulk measurements of concentration. It should also be added that, if a distribution is symmetrically distributed around the mean, the mean, the mode, and the median will be identical numbers.

If we now want to go further and correlate one parameter with another, software analysis packages implement the drawing of two-dimensional plots. Each cell is placed on the plot according to its intensity channel for each of the selected two parameters. Six two-parameter correlations can be derived from our four-parameter data (Fig. 4.2; keep in mind that each two-parameter correlation could be plotted with the x and y axes reversed). Dot plots show, simply, a dot on the page or screen at each locus defining quantitatively (according to channel number) the two relevant characteristics of each particle in the sample. Dot plots suffer, graphically, from black-out in that an area of a display can get no darker than completely black; if the number of particles at a given point are very dense, their visual impact, in comparison with less dense areas, will decrease as greater numbers of particles are displayed.

Contour plots, as another type of two-dimensional graph, display the same kind of correlation as dot plots, but can provide more visual information about the frequency of particles at any given point in the display. They allow the display of data according to the number of cells in any particular cluster (think of the contour lines describing peaks on a mountaineer's map). Lines are assigned to various levels of cell density (as contour maps assign lines to different altitudes) according to any one of several different strategies. While changes in the assignment of lines will not change the values calculated for the number of cells in a cluster, they may radically change the way the data set appears. Figures 4.3 and 4.4 show examples of how the same data plotted with different cell density assignments for contour lines

---➤

Fig. 4.2. The four histograms and six dual-parameter plots derived from the data from a four-parameter cytometer. Each of the six dual-parameter plots could be drawn with the x and y axes reversed.

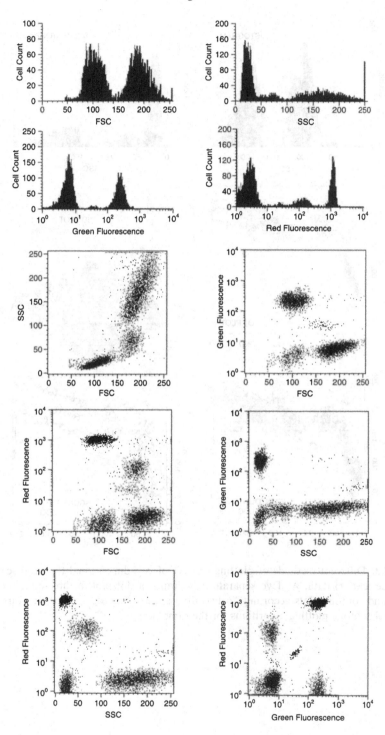

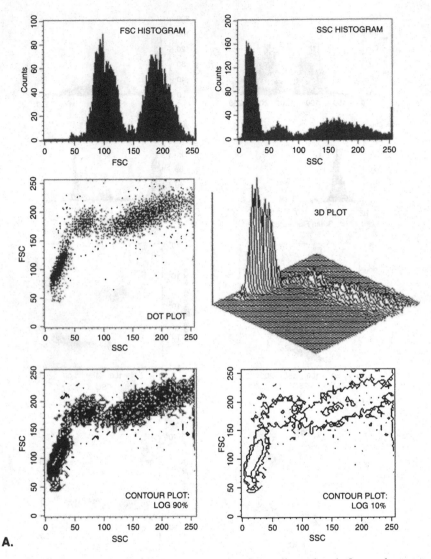

Fig. 4.3. Different methods of plotting one set of two-dimensional (forward scatter vs. side scatter) data. **A:** Two separate histograms, a dot plot, a three-dimensional plot, and contour plots according to two different plotting algorithms. **B:** Four additional contour plotting algorithms for the same data.

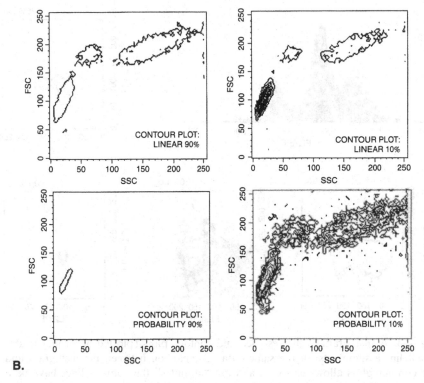

B.

Fig. 4.3 (*continued*)

can make messy data look tidy (or vice versa) and can even make three peaks look like a single distribution. The message here is simply that we need an informed eye when looking at contour plots generated from other people's data.

Once the data from a sample have been plotted in two dimensions, the distribution of particles can then be analyzed. *Quadrants* dividing the plot into four rectangles is the term used for the cursors applied in two-dimensional analysis. As with one-dimensional histograms, these cursors simply divide the light intensity channels into areas of interest. The number of particles in each of the defined areas can then be counted. A standard analysis procedure might use unstained control cells to define the channels delimiting background red and green fluorescence. Quadrants can then be drawn based on these channels,

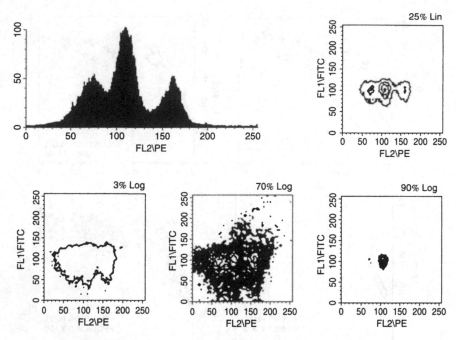

Fig. 4.4. A histogram of fluorescence and contour plots (plotted according to different line assignments) of the same data. Comparison between the histogram and the contour plots allows us to see at what "altitudes" the contour lines have been drawn for each contour plot and why the resulting displays look so different.

and the quadrants will therefore define the staining intensities that we would consider to represent green particles, red particles, and so-called double-positive particles that are both red and green, as well as the double-negative (unstained) cells (Fig. 4.5).

As an intellectual exercise, you should always attempt to visualize the single-dimensional histograms that would result from dot plot or contour plot data. Figure 4.6 indicates dot plots from two different sets of data (with their respective histograms); it is clear that the histograms from these two data sets are identical but that the dot plots are quite different. This example should serve to emphasize that more information is obtained from dual-stained preparations with their two-dimensional plots than from two successive single-dimensional (one color) analyses.

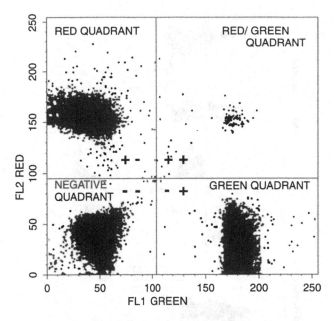

Fig. 4.5. Quadrants are cursors or markers for delineating the intensity (according to channel number) of cells of interest in dual-parameter analysis.

We now need to define *gating*, one of the most subjective aspects of flow cytometry. Gating is based on the defining of *regions*. A region is a way of specifying the characteristics of a subset of particles (in terms of forward scatter, side scatter, and/or fluorescence intensity channel numbers). A region may encompass, for example, all the cells that fall between a certain range of green intensities, or all cells with a certain set of forward and side scatter characteristics, or all cells that are both orange and green. A gate, in contrast, is a combination of regions that define all the cells that we want to include in our final analysis. Cells that pass through the gate get analyzed. A gate can be identical to a single region; that is, the cells we want to include in our final analysis may simply be all the cells that fall into a single, defined region. This is why people often confuse the terms "gate" and "region." However, a gate can also be defined as a combination of two or more individual regions. For example, a gate may include all the cells that fall into region 1 AND into region 2; or

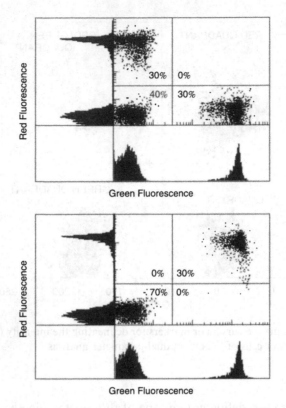

Fig. 4.6. Identical single-color histograms derived from quite different cell samples. The dual-parameter dot plots allow us to distinguish these two populations of cells.

all the cells that fall into region 1 OR into region 2; or all the cells that fall into region 1 but NOT into region 2. If you are young enough, you will have used Boolean algebra in school to combine sets, and these definitions will not present you with any problems. Whether a gate is co-equal to a single region or involves a combination of several regions, any particle that fulfills the defined gate characteristics will pass through the gate and will be included in the next analysis step (Fig. 4.7).

Our gate could, in fact, be a "live gate" or an "analysis gate." A live gate defines the characteristics of cells that need to be fulfilled before the data from those cells are accepted for storage in a computer for further analysis; the information about all other cells will be lost. For this reason, a live gate should be avoided unless data storage

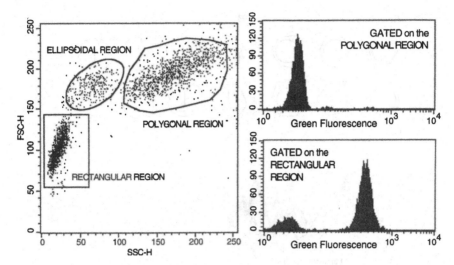

Fig. 4.7. Regions can be drawn to define clusters of cells. Regions can then be used (individually or in Boolean combination) to form gates for restricting subsequent analysis to certain groups of cells.

space is limited. An analysis gate, in contrast, selects cells with defined characteristics from within a large heterogeneous sample of cells, all of whose characteristics have already been stored in a data file. We still have the ability to move the regions around to adapt to different analytical strategies, and we can also calculate the proportion that the gated cells are of the total. The use of either type of gate is rather analogous to the process of, by eye, disregarding all erythrocytes, monocytes, and neutrophils in a microscopic field and then counting only lymphocytes to determine the percentage of lymphocytes that are stained. A trained microscopist's eye is, in that way, defining a lymphocyte gate based on nuclear pattern, shape, and size. A cytometrist can define a lymphocyte's gate in terms of forward and side scatter intensities.

Dual-parameter correlations constitute the standard procedure for analysis of much flow cytometric data. Most cytometers provide us with four or five or more parameters of information. Because humans, in general, are used to thinking in two dimensions, actually correlating three or more parameters with each other can seem rather difficult— in terms of both computer software and our ability to keep track of the strategy (Fig. 4.8). For purposes of data analysis, the most com-

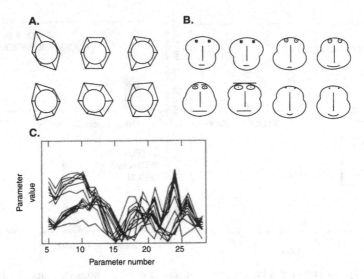

Fig. 4.8. Three different ways to visualize multivariate data. From Dean (1990).

mon procedures use the extra parameters to allow gating (including or excluding) of particles before the final analysis, which is almost invariably just one or two dimensional.

Thus far, we have followed cells into the center of a sheath stream flowing from a nozzle or through a chamber past a focused light beam; we have seen how scatter and fluorescence signals can be generated from those cells when they are hit by that light beam; we have accounted for the registering of those light signals by photodetectors so that the intensities of the signals can be represented by the discrete channels (256 or 1024) of an ADC; we have described the way amplification applied to the output of a photodetector will determine the intensity range represented by the ADC channels; we have seen the way in which the channel number for each of the several parameters characterizing each cell can be stored as list mode data for each cell in sequence in the sample; and we have described the general ways in which the stored data can be displayed and analyzed. We should, therefore, have a clear picture of the fluidic, optical, electronic, and computational characteristics that form the basis for flow cytometric analysis. The chapters that follow will deal with the ways in which these principles are applied to experimental situations. Because the first decisions that are made in designing a flow experiment often concern the specification of reagents for staining cells, we will initially take a short detour to the principles of photochemistry and laser optics in order to understand the requirements that a flow cytometer imposes on these decisions.

FURTHER READING

Chapter 5 in **Shapiro**, Chapter 30 in Volume I of **Weir**, Section 10 in **Current Protocols in Cytometry**, Chapter 7 in **Darzynkiewicz**, and Chapter 22 in **Melamed et al.** are all good discussions of various aspects of flow data analysis. In addition, an entire book by **Watson (1992)** is devoted to this subject.

5

Seeing the Light: Lasers, Fluorochromes, and Filters

Because flow cytometry involves the illumination of particles by a light source and the subsequent analysis of the light emitted by particles after this illumination, an understanding of some of the principles behind light production, light absorption, and light emission is important for the effective design and interpretation of experimental protocols. The principles of photochemistry apply both to the generation of light by a light source and to the absorption and emission of light by a fluorochrome. It is best to get these concepts straight before we begin to describe the staining of cells.

GENERAL THEORY

We need to begin with a brief review of atomic structure. Atoms consist of relatively compact nuclei containing protons and neutrons. At some distance from these dense nuclei each atom has electrons moving in a cloud around the central nucleus. The electrons move in *shells* or *orbitals* or *probability waves* (different words derived from more or less classic or quantum mechanical terms of reference) around the nucleus, and the number of electrons circulating in these orbitals depends on the element in question. Four things are particularly important for flow cytometrists to understand about these electrons: First, atoms have precisely defined orbitals in which electrons may reside. Second, an electron can reside in any one of the defined orbitals but cannot reside in a region that falls between defined orbitals. Third, the energy of an electron is related to the orbital in

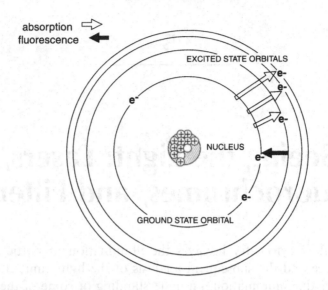

Fig. 5.1. An "old-fashioned" but conceptually easy diagram of an atom with electrons circling the nucleus. Electrons can absorb energy to raise them to an excited state orbital. When they fall back to their ground state, they may emit light, which we call *fluorescence*.

which it resides at any given moment. Fourth, an electron can absorb energy and be pushed to an excited (higher energy) orbital, but it will quickly give back that energy as it rapidly returns to its stable, ground-state configuration (Fig. 5.1). The energy given back as an electron returns from an excited orbital to its ground-state orbital can be in the form of light.

With my apologies to all physical chemists for simplifying electronic structure to four facts, we can now add to our knowledge one more fact about light itself. Light is a form of energy made up of photons, and the color of the light is related to the amount of energy in the photons of that light. For example, when red light is emitted by an object, that object is releasing photons of relatively low energy; blue light, in contrast, is made up of photons of higher energy. *Wavelength* (expressed in nanometers, abbreviated to nm) is a term often used to describe the color of light. The wavelength is inversely related to the amount of energy in the photons of that light: Blue light has wavelengths in the range of 400–500 nm, and its photons have relatively large amounts of energy; red light has wavelengths

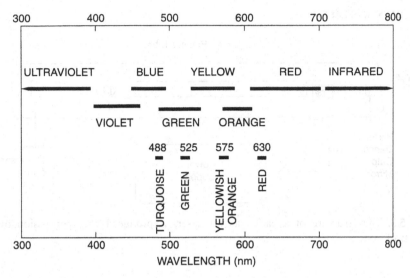

Fig. 5.2. Light, in the region to which our eyes respond, can be described by a "color" or, more precisely, by a wavelength. Photons with longer wavelengths have less energy.

of about 600–650 nm and photons with less energy than those in blue light. The other colors of the visible spectrum fall between these values (Fig. 5.2). Infrared light has a longer wavelength than red light (and less energetic photons); ultraviolet light has a shorter wavelength than blue light (and more energy). With this information, we can start to apply our knowledge of light and atomic structure to flow cytometry.

LASERS

The illumination of particles as they flow past a light source is responsible for generation of the signals upon which flow cytometric analysis is based. The illumination of particles can be provided by an arc lamp or by a laser. In either case, electrons within the light source are raised to high-energy orbitals by the use of electricity, and energy is given off in the form of photons of light when the electrons fall back to their lower energy orbitals. The color of that light is determined by the energy differences between the excited and lower orbi-

THE BASIC ION LASER

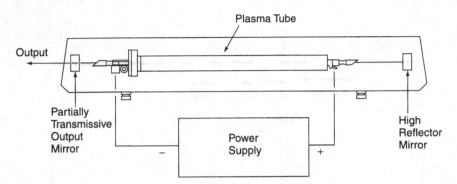

Fig. 5.3. The structure of a basic gas ion laser. Reproduced with permission of Spectra Physics.

tals of the atoms in the lamp or laser. Lasers and arc lamps each have certain advantages. Because lasers are the light source of choice in most current flow systems, it is important to understand the definite restrictions on experimental flexibility that are imposed by the way in which a laser works.

Gas lasers consist of tubes (called *plasma tubes*) filled with gas; a cathode lies at one end and an anode at the other (Fig. 5.3). A voltage is applied across the plasma tube in order to raise the electrons in the atoms of gas to excited orbitals. As the electrons fall back to lower energy ground states, they give off energy in the form of photons of light; the color of the emitted light is determined by the type of gas used and is a function of the differences between the energy levels of its atomic orbitals. Reflecting mirrors are placed outside the tube at either end. If the reflecting mirrors are aligned precisely with the plasma tube, then the photons given off by the gas will be reflected back and forth through the tube between the two reflecting mirrors. The applied voltage maintains the electrons in the gas in excited orbitals. The reflected photons, as they oscillate back and forth within the tube, interact with these excited gas ions to stimulate them to release more photons of identical energy (in a way predicted by Einstein). An amplification system thereby results: The photons oscillating back and forth between the two end mirrors cause more and more light to join the beam. By allowing a small percentage of this oscillating beam to leave the system at the front mirror, we have generated what is known as a *coherent* light source.

A laser light source is coherent in three respects. It is coherent as regards its direction in that a beam is generated that diverges little and remains compact and bright for great distances (as in laser light shows and "star wars"). It is coherent with respect to polarization plane (of possible relevance for some specialized aspects of cytometry). Finally, it is coherent with respect to color because the electrons in a gas atom or ion are restricted in the orbitals that they use under plasma tube conditions and are therefore restricted in the amount of energy that is emitted when they fall from a higher energy to a lower energy state.

Coherence with respect to direction is the reason lasers are useful in flow cytometry: They provide a very bright, narrow beam of light allowing particles flowing in a stream to be illuminated strongly for a very short period of time so that measurable signals from one particle can be generated and then separated by darkness from signals generated by the following particle (refer back to Figs. 3.1 and 3.5). The disadvantage of this spatial coherence is that cells in a stream must be well aligned in the center of that stream if they are going to be uniformly illuminated.

Although spectral purity has certain advantages, coherence with respect to color is actually one of the major limitations of a laser system. Whereas an ordinary light bulb will put out light of a wide range of colors (a white light bulb puts out a mixture of the whole range of colors in the visible spectrum), the color of the output from a laser is restricted by the restricted range of electron orbitals that will support lasing in any particular gas. Argon ion lasers are the most common lasers used in flow cytometry today; they emit useful amounts of light at 488 nm (turquoise) and at 514 nm (green), as well as small amounts of ultraviolet light. By using a prism or a wavelength-selective mirror, the operator or manufacturer can choose one or the other of these colors, but there is relatively little light available in other regions of the spectrum (Fig. 5.4). Helium-neon lasers or red diode (solid-state) lasers are also used (often in conjunction with an argon ion laser in a dual laser benchtop flow system); a helium-neon laser puts out red light at 633 nm and a red diode laser at 635 nm. Research instruments may have three lasers, with one or two argon lasers for 488 nm and ultraviolet excitation plus a third HeNe laser

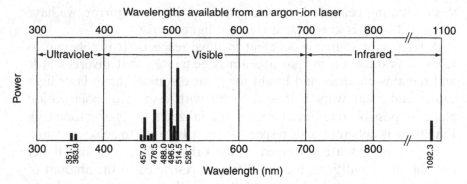

Fig. 5.4. The wavelengths of light emitted by an argon ion laser. The most powerful (and therefore most useful) wavelengths are 488 and 514 nm. High-power argon ion lasers can provide some useful light in the ultraviolet range.

or dye laser for red excitation. As a general principle, flow cytometers all have argon lasers emitting light at 488 nm; more advanced instruments will add additional lasers with additional wavelengths as required (Table 5.1).

A laser is therefore very good at providing bright, narrow, stable

TABLE 5.1. Laser Options with Examples of Some Possible Fluorochromes

Laser type	Wavelengths (nm)	Possible fluorochromes
Argon ion	488	Fluorescein, R-phycoerythrin, PerCP, PE-tandems, EGFP, EYFP, propidium iodide, Alexa 488, acridine orange
	514	Rhodamine, propidium iodide, EYFP, R-phycoerythrin
	Ultraviolet (351/364)	Hoechst dyes, DAPI, Indo-1, AMCA, Cascade Blue, EBFP
Helium-neon (HeNe)	Usually 633	Allophycocyanin, Cy5, TO-PRO-3
Krypton ion	568	Cy3, Texas Red, Alexa 568
	647	Allophycocyanin, Cy5, TOPRO-3
Red diode	635	Allophycocyanin, Cy5, TOPRO-3
Helium-cadmium (HeCd)	325	Indo-1, propidium iodide
	442	Mithramycin, chromomycin A3, ECFP

light beams of well-defined color and intensity. It is inflexible, however, in terms of the color of that light. In addition, large lasers are demanding in their requirement for routine, skilled maintenance; careful alignment between plasma tube, reflecting mirrors, output beam, and stream is essential. Furthermore, lasers are expensive; plasma tubes have limited lives and cost considerably more than an arc lamp when they need replacing. (It is worth noting that a well-aligned plasma tube with clean mirrors will last longer than a poorly aligned and dirty one because it will draw less current to produce the same amount of light; loving care of a laser can make a considerable difference to the overall running costs of a flow cytometer.) Many flow systems use small, air-cooled lasers. Air cooling is feasible when the laser is of low power and does not generate much heat (gas lasers are all relatively inefficient and generate a good deal of waste heat). However, cytometers with a stream-in-air analysis point have space and refractive index–mismatch surfaces between the cells and the light collecting lenses; as a result, detection of the light signals generated by the cells is inefficient. For this reason, sorting cytometers may require large, high-power lasers that put out very bright beams of light but also generate a great deal of waste heat. These sorting cytometers usually need a large supply of cold water to keep their lasers cool and also need skilled cleaning and alignment. The main fact to be kept in mind about lasers, as we move to a discussion of fluorochromes, is that, because lasers make use of electrons excited to a limited range of orbitals, they are restricted in the color of the light they emit.

FLUOROCHROMES

In a laser, electrons are excited by means of energy from an applied voltage. We then take advantage of the energy given off as light when the electrons fall back to lower energy states. With a fluorescent molecule (a "fluorochrome" or "fluorophore"), light itself (called the *excitation light*) is used to excite the electrons initially. We then analyze the emitted light (of a different color) that is given off as the electrons of the fluorochrome return to their ground-state orbitals. On the basis of our knowledge about electron orbitals together with our knowledge about laser output, it should now be apparent that, when a fluorochrome is illuminated, the electrons in that fluorochrome

can absorb energy if, and only if, the excitation light contains photons with just the right amount of energy to push an electron from
one defined fluorochrome orbital to a higher (more energetic) defined
orbital. Thus the difference between orbital energy levels of a fluorochrome will strictly determine what color of light that particular
fluorochrome can absorb. When electrons fall back from their excited
(more energetic) levels to their ground-state or less energetic orbitals,
light is emitted of a color that depends on the difference in energy
levels between the two fluorochrome orbitals in question. Because
energy is not created and is, in fact, always lost to some extent as
heat, the color of the light emitted when an electron falls back from
an excited state to its ground state is always of a somewhat lower
energy (that is, of a longer wavelength) than the energy absorbed in
raising the electron to the higher orbital in the first place (Fig. 5.5).
The light given off when an electron falls back from an excited state
to its ground state orbital is called *fluorescence*. The time required
for fluorescence to take place is approximately 10 ns after the initial
activation of the fluorochrome by the excitation beam.

If we plot the colors of light that can be absorbed and the colors
that are then emitted by various compounds (called *absorption* and

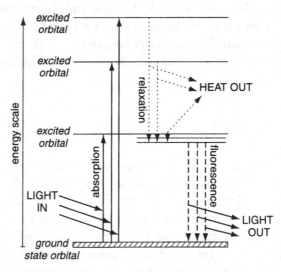

Fig. 5.5. The absorption of light by an electron and the subsequent emission of the
energy from that light as both light and heat. Because some of the absorbed energy is
lost as heat, the emitted light always has less energy (and is of longer wavelength)
than the absorbed light.

emission spectra), we can see how these principles work in practice (Fig. 5.6). The wavelengths of light absorbed by a compound will depend exactly on the electronic orbitals of its constituent atoms; the light then given off as fluorescence is always of a longer wavelength than the absorbed light. The difference between the peak wavelength for absorption and the peak wavelength for fluorescence emission is known as the *Stokes shift*. Some compounds (for example, PerCP) have a larger Stokes shift than others (for example, fluorescein).

We are now in a position to understand why the use of a laser to provide the illuminating beam in a flow system restricts the choice of fluorochromes that can be used for staining cells. If we are using an argon ion laser with an output of light at 488 nm, we can consider as suitable stains those and only those fluorochromes that absorb light at 488 nm. Rhodamine, a stain used extensively by microscopists, absorbs light poorly at 488 nm and is therefore not useful in conjunction with a 488 nm laser. Stains like DAPI and Hoechst can be used with a high-energy argon ion laser tuned to its ultraviolet line, but cannot be used if the laser is tuned to 488 nm. Appropriate stains for 488 nm excitation include $DiOC_n(3)$ for looking at membrane potential and propidium iodide and acridine orange for looking at nucleic acid content. With the 488 nm light from an argon laser, the situation is also ideal for staining cells with fluorescein (another standby of microscopists). In fact, this traditional allegiance to fluorescein (sometimes abbreviated as FITC; fluorescein isothiocyanate is the chemically active form of the dye that will conjugate to proteins) is the principal reason that argon ion lasers were initially selected for the first laser-based flow cytometers. Fluorescein absorbs light in the range of 460–510 nm and then fluoresces in the range of 510–560 nm, with a peak at about 530 nm (green); it can also be readily conjugated to antibodies, thereby providing specific fluorescent probes for cell antigens (Table 5.2).

Because multiple photodetectors are available, a flow cytometer has the ability to measure two or more fluorescence signals simultaneously from the same cell. To use several fluorochromes at the same time, cytometrists with only one laser required a group of stains, all of which absorb 488 nm light but which have different Stokes shifts so that they emit fluorescent light at different wavelengths and thereby can be distinguished from each other by the color of their fluorescence. Propidium iodide and fluorescein are a pair of fluorochromes that fulfill these criteria (having different Stokes shifts) and can be

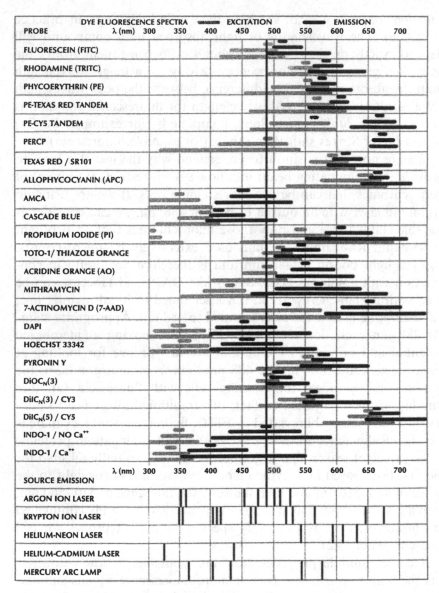

Fig. 5.6. The absorption (gray bands) and emission (black bands) spectra of various fluorochromes. The wavelength widths of the bands for each fluorochrome indicate the range of wavelengths that will be absorbed and emitted. Laser (excitation) wavelengths are indicated at the bottom of the chart. From Shapiro (1995).

TABLE 5.2. **Absorption and Emission Wavelength Maxima
of Some Useful Fluorochromes**[a]

Fluorochrome	Absorption (nm)	Emission (nm)
Covalent labeling of proteins		
Fluorescein (FITC)	492	520–530
R-phycoerythrin (PE)	480–565	575–585
PerCP	490	677
Alexa 488	494	519
PE–Texas Red tandem	480–565	620
PE–Cy5 tandem	480–565	670
PE–Cy7 tandem	480–565	755
Tetramethylrhodamine (TRITC)	540	570
Cy3	540	567
Alexa 568	578	603
Texas Red	595	620
Cy5	648	670
Allophycocyanin	650	660
Nucleic acids		
Hoechst 33342 or 33258	346	460
DAPI	359	461
Chromomycin A3	445	575
Propidium iodide	480–580 (also UV)	623
Acridine orange		
+ DNA	480	520
+ RNA	440–470	650
Thiazole orange	509	525
7-aminoactinomycin D (7-AAD)	555	655
TO-PRO 3	642	661
Membrane potential		
$diOC_n(3)$ oxacarbocyanines	485	505
Rhodamine 123	511	546
pH		
Carboxyfluorescein		
High pH	495	520 (brt)
Low pH	450	520 (wk)
BCECF		
High pH	508	531 (brt)
Low pH	460–490	531 (wk)
Calcium		
Indo-1		
Low calcium	361	485
High calcium	330	405

(continued on next page)

TABLE 5.2 (*continued*)

Fluorochrome	Absorption (nm)	Emission (nm)
Fluo-3		
High calcium	490	530 (brt)
Low calcium	490	530 (wk)
Reporter molecules		
E (enhanced) GFP	489	508
EBFP	380	440
ECFP	434	477
EYFP	514	527
DsRed	558	583
FDG (β-galactosidase substrate)	490	525
Cell tracking dyes		
CFSE	495	519
PKH 67	485	500
PKH 26	510 and 551	567

[a] The absorption and emission maxima from this table will provide clues to the spectral ranges that are useful for excitation and for fluorescence detection with a particular fluorochrome. However, the absorption and emission spectra have breadth, with slopes and shoulders and secondary peaks (see Fig. 5.6). With efficient fluorochromes, excitation and fluorescence detection at wavelengths distant from the maxima may be possible. Therefore, inspection of the full, detailed spectra is necessary to get the full story. In addition, spectra may shift in different chemical environments (this will explain why maxima vary in different reference tables from different sources). Values in this table are derived primarily from the **Molecular Probes Handbook** and the article by Alan Waggoner (Chapter 12) in **Melamed et al.**

used together for analyzing nucleic acid content in fluorescein-labeled cells. Propidium iodide, however, is not fluorescent unless bound to double-stranded nucleic acid and therefore is not usefully conjugated to antibodies for the staining of cell surfaces. Flow cytometrists were therefore spurred to hunt for appropriate pigments (absorbing 488 nm light, fluorescing at a wavelength different from 530 nm, and capable of chemical conjugation to proteins). The humble seaweeds provided some of the answers.

Algae need to absorb light for photosynthetic metabolism, but many of them live in dimly lit regions beneath the surface of the ocean. In this inhospitable marine environment, they have evolved a rich assortment of pigments to help them capture light efficiently. Surveying this array of algal pigments, we can find compounds that absorb light and fluoresce at a range of different wavelengths. A quick inspection of the absorption and fluorescence emission spectra (Fig. 5.6) indicates that, with a light source of 488 nm, phycoerythrin will absorb light fairly efficiently. With its longish Stokes shift, it fluoresces around 578 nm (orange). In addition, phycoerythrin was found to be almost as suitable as fluorescein for conjugating to antibodies or other proteins. For these reasons, fluorescein and phycoerythrin (PE) have become the fluorochromes of choice for dual color flow cytometry. They can both be conjugated to antibodies to provide specific reagents that fluoresce with different colors.

More recently, other algal pigments have been developed for flow cytometry. The dinoflagellate *Peridinium* has a photosynthetic apparatus that includes complexes of eight carotenoid ("peridinin") molecules surrounding two chlorophyll molecules. This carotenoid: chlorophyll complex is called *peridinin chlorophyll protein* (marketed as PerCP by Becton Dickinson). The carotenoid molecules absorb blue-green light, and, because of the proximity of the chlorophyll molecules, the carotenoids will pass the absorbed energy (by a process called *nonradiative* or *resonance energy transfer*, i.e., no light is emitted) on to the adjacent chlorophylls. The chlorophylls, now with electrons in an excited state, can use this energy either for photosynthesis (in the alga) or for fluorescence (in the flow cytometer). In this way, the complex couples the absorption properties of one molecule with the emission properties of another molecule, shifting the wavelength of fluorescence from that of the primary fluorochrome to that of the secondary fluorochrome—thus effectively increasing the Stokes shift. PerCP can thus be used as a third color in conjunction with fluorescein and phycoerythrin.

In imitation of nature, organic chemists have engineered "tandem" fluorochromes that mimic the light-harvesting complexes of the algae. These synthetic pigments consist of two different fluorochromes bound together on a single backbone so that the first fluorochrome absorbs light from the laser and passes the energy on to the second fluorochrome, which emits the light at a relatively long wavelength. A common example of a synthetic tandem fluorochrome consists of a phycoerythrin molecule covalently linked to a Texas Red molecule or to a Cy5 molecule. In the presence of 488 nm light, the phycoerythrin moiety will absorb light, but, rather than fluorescing itself, will pass the absorbed energy on to the closely linked Texas Red or Cy5 molecule, which will then fluoresce at its own wavelength beyond 600 nm. These tandem molecules (whether natural or synthetic) provide an ideal way to increase the multicolor staining options when only a single 488 nm laser is available. The synthetic tandems, in particular, do have certain intrinsic problems: The efficiency of light transfer between the primary and secondary fluorochromes of the tandem can vary depending on manufacturing procedures and on storage. This means that both the intensity and the color of its fluorescence can vary from one batch to another and over time. Although a common combination of fluorochromes for three-color analysis using 488 nm excitation is fluorescein, phycoerythrin, and a tandem of PE with Cy5, the use of these tandem fluorochromes can be avoided if more than one laser is available for excitation.

To increase options further within the limitations of the available fluorochromes, cytometrists are forced to purchase additional lasers. An additional laser will be focused to strike the stream at a secondary analysis point, usually downstream from the 488 nm illumination spot. By the use of a helium-neon or red diode laser, emitting light at 633–635 nm, fluorochromes such as allophycocyanin or Cy5 can be used. Allophycocyanin (APC) is a natural algal pigment that is a single molecule, not a tandem complex. Cy5 is a synthetic cyanine dye that can be modified to alter its absorption and emission characteristics (see Cy3 and Cy7). In addition, Molecular Probes (Eugene, OR) have developed an attractive set of fluorochromes (the Alexa dyes) that span the range of spectral properties suitable for various lasers. A cytometer with two argon ion lasers or with an argon ion and a helium-cadmium (He-Cd) laser can provide the option of using ultraviolet light at the same time as 488 nm light. This also increases the range of fluorochromes that can be analyzed. Research cytometers now are usually adapted for the implementation of two- or three-laser illumination. Increasingly, routine clinical cytometers have also begun to contain a second laser on their optical benches.

PARTITIONING THE SIGNAL WITH LENSES, FILTERS, AND MIRRORS

We have now described a system in which one or more narrow beams of laser light of well-defined wavelength are used to illuminate a cell, and the light scattered by the cell and emitted by various fluorochromes in or on that cell provide signals that are registered on a group of photodetectors. From our description of the optical bench in Chapter 3 (refer back to Fig. 3.7), we should recall that there are three, four, or more detectors whose job it is to measure the intensity of the light coming out at right angles from the cells in the illuminating beam. The light coming out at right angles will be a mixture of colors, depending on the fluorochromes with which the cell has been stained. Because photodetectors are placed at different spatial positions in the cytometer, we need to use a combination of mirrors to direct light of a specific color toward its intended photodetector and not to any other (Fig. 5.7). Recalling that photodetectors are relatively color blind, it is these mirrors combined with colored filters that give each photodetector its own spectral specificity.

An optical mirror is vacuum coated precisely with thin layers of metal so that it will split a light beam by reflecting some of the light shining onto its surface while transmitting the rest. Mirrors that split a light beam according to color are called *dichroic mirrors*. A 640 nm short-pass dichroic mirror (640 SP), for example, if placed at a 45° angle in a light path, will transmit light of all wavelengths less than 640 nm and reflect all wavelengths greater than 640 nm to the side. In contrast, a 500 nm long-pass dichroic mirror (500 LP) will transmit all light of wavelengths greater than 500 nm, but reflect to the side all light less than 500 nm. (These dichroic mirrors are angle sensitive; they are usually used at 45° to the light path. If the angle is changed, the wavelengths of the transmitted and reflected light will also change.)

By following the light path in Figure 5.7, we can see the way dichroic mirrors (in an example of a typical two-scatter parameter plus three-fluorescence parameter system) can be used to partition multicolor light. The first dichroic mirror (a 500 nm LP) reflects all light less than 500 nm to the side toward one photomultiplier tube (PMT); all light with wavelengths greater than 500 nm is transmitted straight ahead. At the next dichroic mirror, a further partition takes place: All light with wavelengths greater than 640 nm is reflected to the side from this 640 nm SP mirror, and light less than 640 nm is transmitted. Because light less than 500 nm has already been removed from this beam, the transmitted light at this stage will be of wavelengths between 500 and 640 nm. At the final dichroic mirror in this system, light greater than 560 nm will be reflected out of the beam (because of the previous step, this light will be between 560 and 640 nm), and all light shorter than 560 nm will continue straight ahead (being between 500 and 560 nm because of previous partition steps). By use of a system such as this:

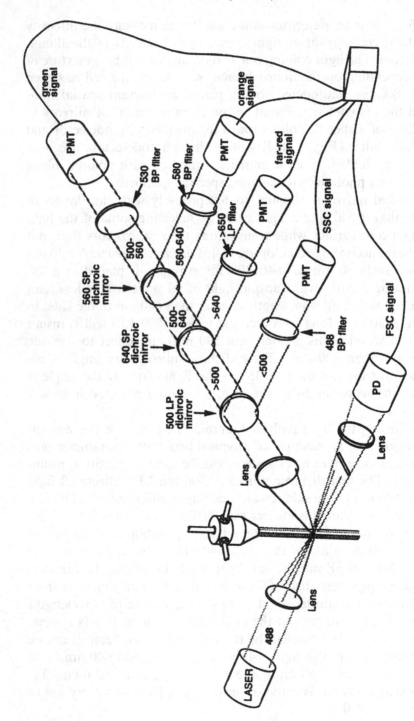

Fig. 5.7. A filter and mirror configuration for making five photodetectors specific for registering forward scatter, side scatter, green, orange, and far-red signals.

· Side scatter light (488 nm) will be reflected out of the beam path toward the first PMT by the 500 LP mirror
· Far-red fluorescence will be reflected out of the beam path toward the second PMT by the 640 SP mirror
· Orange fluorescence will be reflected out of the beam path toward the third PMT by the final 560 SP mirror
· Green fluorescence will be transmitted straight ahead through that final mirror toward the last PMT in the optical system

By the use of filters with particular color specifications immediately in front of the photodetectors, we have our last chance for making sure that the only light registered on a given photodetector has been derived from a specific fluorescence color. Filters are similar to dichroic mirrors, but are meant to be used at 90° to the light path; they transmit light according to its color. Filters come in three basic types: A short-pass filter will transmit all light of a wavelength less than the specified value; a long-pass filter will transmit all light of a wavelength greater than the specified value; and a band-pass filter will transmit only light in a narrow wavelength band around the specified value. Band-pass filters are of most use in flow systems because they are most specific in their transmission characteristics. These characteristics are usually described according to the peak transmission wavelength and the width of the wavelength band at 50% of that peak transmission. For example, if a filter transmits 90% of the 570 nm light falling onto its surface, and if this percentage of transmission drops to 45% when the wavelength changes by 15 nm in either direction, then the filter will be described as a 570/30 nm band-pass filter.

Different filter and mirror combinations will be required for experiments using different combinations of stains. By looking at the detailed fluorescence emission spectra of the fluorochromes in use, we can plan a strategy for selecting the filter and mirror sets that will allow us to correlate the signal from a given photodetector with the fluorescence from a given fluorochrome. Figure 5.8 shows the fluorescence emission spectra from fluorescein and phycoerythrin drawn on the same axes. (These two fluorochromes are the most useful example for our purposes here, but the same principles and filter strategy would apply to any group of two or more fluorochromes used simultaneously.) By inspection of these spectra, we can see that the best dichroic mirror to use for splitting fluorescein fluorescence from phycoerythrin fluorescence is a 560 nm mirror. This will send most of

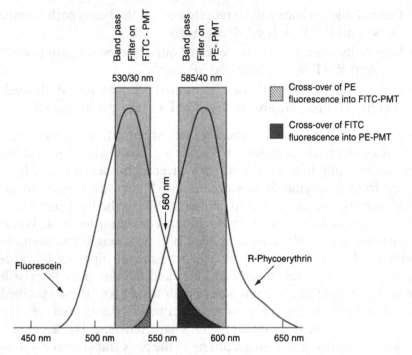

Fig. 5.8. The cross-over of fluorescein and phycoerythrin fluorescence through the filters on the "wrong" photodetectors.

the fluorescein fluorescence straight through to one photodetector and reflect most of the phycoerythrin fluorescence toward a different photodetector. By fitting the green detector with a 530/30 nm filter and the orange detector with a 585/40 nm filter, we have constructed a system that will register fluorescein fluorescence and phycoerythrin fluorescence mainly on different photodetectors. We might be entitled now to call one photodetector the "fluorescein detector" and the other one the "phycoerythrin detector"; but that qualifying word *mainly* lies at the root of our next problem.

SPECTRAL COMPENSATION

With the filter and mirror combinations just described, a cell stained only with fluorescein will emit fluorescence that will register strongly on the green-specific photodetector (and this is fine if we are using only fluorescein as the stain in our system). However, inspection of the fluorescence emission spectrum of fluorescein (Fig. 5.8) indicates that some of the light emitted by a cell or other particle stained only with fluorescein will manage to evade our strategic obstacles (mirrors and filters); some of the light emitted by fluorescein is of a wavelength long enough to register on the orange photodetector. This is another way of saying that the light emitted by fluorescein is "orange-ish green." No matter how tightly we restrict the filter and mirror specifications, we cannot get round the fact that there is a significant degree of overlap between the fluorescence emission of fluorescein and that of phycoerythrin. Therefore, there is no way to keep all the fluorescein-derived light from the orange detector without losing most of the phycoerythrin fluorescence as well. If we want to measure PE fluorescence with reasonable sensitivity, then the orange photodetector will also respond to a limited extent to fluorescein fluorescence. Our use of filters and mirrors must, for these reasons, be a compromise.

What this compromise means, in practice, is that a cell stained only with fluorescein will register a signal brightly on the green-specific photodetector but also significantly on the orange-specific photodetector. Similarly, a cell stained only with phycoerythrin will register a signal brightly on the orange detector, but also slightly on the green detector. The left panel of Figure 5.9 shows how these pure signals look on dot plots giving the relative intensities for the signals on the orange and green PMTs. Clearly, we would have been incorrect in calling the detectors fluorescein or phycoerythrin specific; the filters and mirrors in front of our photodetectors cannot make them as specific as we would like them to be. The correct terminology should, therefore, simply identify a photodetector by the specifications of its filter (that is, a "530/30 nm detector" or a "585/40 nm detector"). When flow cytometrists, being as lazy as everyone else, talk about a "fluorescein photodetector," you now have enough knowledge to understand the imprecision of this phrase. In many cytometers, the fluorescence detectors are simply given numbers (e.g., FL1, FL2, FL3). If the filters are fixed (as in some benchtop instruments), the user needs to refer to the technical manual to know which filters are in front of which detectors. In a research cytometer, the filters can be changed to adapt to the use of different combinations of fluorochromes; knowledge of the exact optical pathway in place at any one time becomes correspondingly much more important.

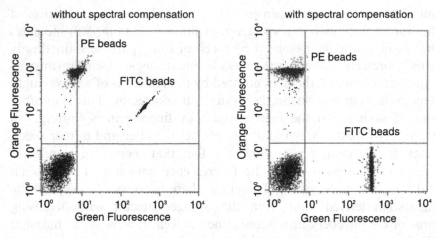

Fig. 5.9. The fluorescence from phycoerythrin beads registers a bit on the green photodetector; and the fluorescence from fluorescein beads registers considerably on the orange photodetector. Compensation (right) allows us to correct for this cross-over.

Having understood why there is cross-over or overlap between fluorochromes, each registering to a certain extent on the "wrong" photodetector, we now must decide what to do about it. What we do about it is called *compensation.* Cytometers are provided with an electronic circuit, a compensation network, that measures the intensity of the signal on one photodetector and subtracts a certain percentage of that signal from the signal on another photodetector (because a certain fixed percentage of any fluorescein signal will always cross over and register on the 585/40 nm phototube, and a certain fixed percentage of any phycoerythrin signal will always cross over and register on the 530/30 nm phototube).

Those percentages are routinely determined empirically. Cells brightly stained with only fluorescein are run through the cytometer; their signals on both the green and orange photodetectors are recorded; and the percentage subtraction (percentage of the signal on the green detector to be subtracted from any signal on the orange detector) is varied until no signal above background is registered on the orange photodetector. Technically, we want to make sure that the median orange channel of fluorescein-stained cells is identical to the median orange channel of unstained cells. We then go through the same procedure with cells that have been brightly stained only with phycoerythrin in order to determine what percentage subtraction is required to compensate the green photodetector for cross-over from the phycoerythrin fluorescence. When both photodetectors have been correctly compensated for cross-over from the wrong fluorochromes, our signals, plotted as two-dimensional dot plots, should look like those in the right panel of Figure 5.9.

If more than two fluorochromes have been used for simultaneous multiparameter analysis, then the signals from all PMTs need to be compensated individually in the same way with respect to each fluorochrome (i.e., in a three-color example, the green PMT needs to be compensated for cross-over fluorescence from both the red and orange fluorochromes; the orange PMT for cross-over fluorescence from the green and red fluorochromes; and the red PMT for cross-over fluorescence from the green and orange fluorochromes). In practice, this involves staining cells separately and brightly with each of the three fluorochromes. The separately stained cells can then each be mixed with unstained cells; the three mixtures can be run through the cytometer and examined with respect to three different dot plots (red vs. green; green vs. orange; and red vs. orange) to make sure that the compensation network gives square patterns (as in Fig. 5.9, right panel) for the cells in all three plots.

As a general comment, compensation is one of the more difficult areas in flow cytometry and gets more difficult as more colors are involved. For this reason, there is software available that permits postexperiment compensation: Samples can be run through the cytometer without any or with minimal electronic compensation, and the compensation subtraction can be applied or adjusted afterward during data analysis. There are great advantages to software compensation because, especially in multicolor experiments, it is often difficult to "get it right" while under the pressure of acquiring data at the cytometer bench. Of course, even postexperiment compensation still requires that files from single-stained samples be stored so that the software compensation subtraction can be evaluated.

It remains to be said (unfortunately) that the compensation values are valid only for a particular pair of fluorochromes with a particular set of filters and mirrors and with particular voltages applied to a particular set of PMTs. If any one of those elements is altered, the required compensation values will alter as well. In general, once compensation has been set using single-stained particles with a given experimental protocol (PMT voltages, filters, and so forth), compensation values between a given pair of fluorochromes should remain constant within that protocol from day to day. However, because compensation can affect the way two-color plots are evaluated, correct compensation can be of critical importance in the interpretation of data from samples that are brightly stained with one fluorochrome but weakly stained with another. For this reason, single-stained controls should always be run within each experimental group in applications where this type of interpretation is required.

Through the chapters so far, a basic cytometric system has been described. We have generated an illuminating beam, followed the fluid controls that send cells through the center of that beam, and registered the light signals emerging from those cells as specifically as possible on individual photodetectors, according to their color. We have also compensated for any lack of specificity in our photodetectors that might result from spectral cross-over between fluorochromes. Finally, we have stored the information from the light signals from each cell in a computer storage system so that they can be analyzed in relation to each other through computer software. We are now in a position to begin to look at ways in which such a flow system might be used. In the following chapters, applications will be described to give some indication of the common, the varied, and the imaginative practices of flow cytometry. My choice of specific examples is intended to illustrate points that may be of general importance and also to illustrate a range of applications that might stimulate interest in readers who are new to the field. It is not meant, nor in a short book could it hope, to be exhaustive.

FURTHER READING

Chapter 2 in **Ormerod** is a good, gentle discussion of fluorescence technology and light detection. Chapter 4 in **Shapiro** is slightly more detailed, and Chapter 2 in **Van Dilla et al.** follows well from there. The **Melles-Griot** catalog is a valuable (and free) source of information about lasers and filters.

Chapter 12 in **Melamed et al.**, Section 4 in **Current Protocols in Cytometry**, and Chapter 7 in **Shapiro** discuss the uses of different fluorochromes. With respect to the physical characteristics of these fluorochromes, the **Molecular Probes Handbook** is a valuable (and free) source of information. Chapter 1.14 in **Current Protocols in Cytometry** is a detailed discussion by Mario Roederer of compensation.

6

Cells from Without: Leukocytes, Surface Proteins, and the Strategy of Gating

In this chapter, the analysis of leukocytes, one of the more important applications of flow cytometry, is discussed. It is an important application not because leukocytes are intrinsically more important than other types of cells but because they constitute a class of particles that, for several reasons, are ideally suited for analysis by flow cytometry and thus make very good use of the capabilities of the technique. Leukocytes therefore account for a large proportion of the material analyzed by flow cytometry in both hospital and research laboratories. They also serve as good examples for describing here some of the general procedures of cytometric protocol, particularly the art of gating and the requirements for controls.

Leukocytes are well suited to flow analysis for three reasons. First, they occur in vivo as single cells in suspension; they can therefore be analyzed without disaggregation from a tissue mass (and no spatial information is lost in the course of preparation). Second, the normal suspension of leukocytes (i.e., blood) contains a mixture of cell types that happen to give off forward (FSC) and side (SSC) light scatter signals of different intensities, thereby allowing them to be distinguished from each other by flow cytometric light scatter parameters. Third, with monoclonal antibody technology, a vast array of specific stains for probing leukocyte surface proteins has been developed; these stains have allowed the classification of microscopically identical cells into various subpopulations according to staining characteristics that are distinguishable by means of flow analysis.

I will first describe leukocytes in a general way for those not familiar with this family of particles. We will then proceed to a description of general staining techniques for surface markers, with attention to the need for controls. At this point I will digress slightly and discuss the way cytometry results can (and cannot) be quantified and the way sensitivity can affect these results. Finally, I will discuss gating strategies, in both philosophy and practice.

CELLS FROM BLOOD

Whole blood consists of cells in suspension (Fig. 6.1). In a "normal" milliliter (cm^3) of blood, there are about 5×10^9 erythrocytes (red blood cells), which are shaped like flat disks (about 8 μm in diameter) and are responsible for oxygen transport around the body. In addition, per cm^3 there are about 7×10^6 white cells (leukocytes) that collectively are involved in immune responses in the organism. The leukocytes appear to be heterogeneous under the microscope and can be divided according to their anatomy into several classes. There are cells that are called *monocytes*, which are of a concentration of about 0.5×10^6 per cm^3; these appear on slides as round cells (about 12 μm) with horseshoe-shaped nuclei and are responsible for phagocytosis of invading organisms and for presenting antigens in a way that can initiate the immune response.

There are also cells that appear as 10 μm circles with large round nuclei and a narrow rim of cytoplasm; these are the *lymphocytes*, present in a concentration of about 2×10^6 per cm^3. Lymphocytes (including both B lymphocytes and T lymphocytes) are responsible for a great variety of the functions that are known as *cellular* and *antibody* (humoral) *immunity*, as well as for the regulation of those functions. The third (and most frequent) group of leukocytes occurs

---

Fig. 6.1. Scanning electron micrographs showing the different surface textures of red (Er) and white blood cells. **A:** Cells within a blood vessel. **B,C:** A comparison of scanning electron micrographs with conventional light microscope images of the same field of stained cells. Enlarged pictures at the right emphasize the different surface textures of monocytes (Mo) and platelets (Pl) in **D,** lymphocytes (Ly) in **E,** and neutrophils (Ne) in **F.** From Kessel RG and Kardon RH (1979). Tissues and Organs: A Text Atlas of Scanning Electron Microscopy, WH Freeman, NY.

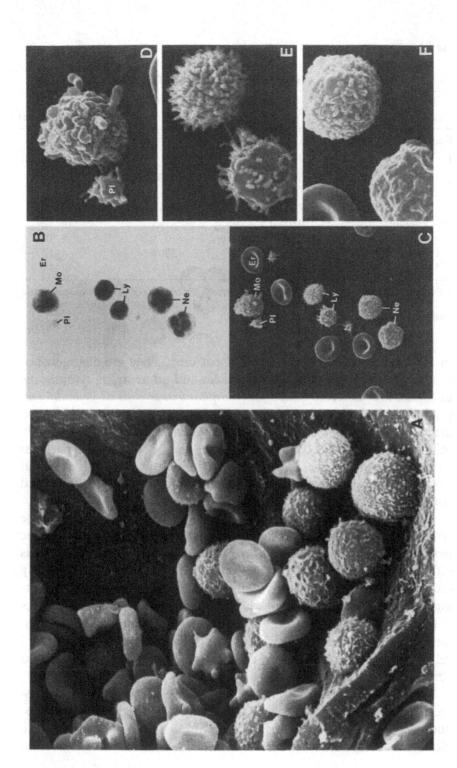

TABLE 6.1. Cells in Normal Human Adult Peripheral Blood

Cells	No. per cm^3 (95% range)	Percent of WBC	Diameter (μm)
Platelets (thrombocytes)	1–3 × 10^8		2–3
Erythrocytes (RBC)	4–6 × 10^9		6–8
Leukocytes (WBC)	3–10 × 10^6	[100]	
Granulocytes			
Neutrophils	2–7 × 10^6	50–70	10–12
Eosinophils	0.01–0.5 × 10^6	1–3	10–12
Basophils	0–0.1 × 10^6	0–1	8–10
Lymphocytes	1–4 × 10^6	20–40	6–12
Monocytes	0.2–1.0 × 10^6	1–6	12–15

Values are from Lentner C, ed. (1984). Geigy Scientific Tables, 8th edition. CIBA-Geigy, Basle; and from Diggs LW, et al. (1970). The Morphology of Human Blood Cells, 5th edition. Abbott Laboratories, Abbott Park, IL.

in a concentration of about 5 × 10^6 per cm^3. They are distinguished under the microscope by lobular nuclei and an array of cytoplasmic granules, are therefore called *polymorphonuclear cells* or *granulocytes* (including basophils and eosinophils, but mainly neutrophils), and are collectively responsible for phagocytic as well as other immune-related activities. In addition, there are small, cell-derived particles called *platelets* (about 2 μm in size), in a concentration of about 2 × 10^8 per cm^3. The platelets are involved in mechanisms for blood coagulation (Table 6.1).

Although the blood is an easily accessible tissue to study, because red blood cells outnumber leukocytes by about 1000 to 1 in the peripheral circulation, the analysis of leukocytes by any technique is difficult unless the red cells can be removed. Techniques for removing red cells usually involve either density gradient centrifugation (pelleting red cells and neutrophils, leaving lymphocytes and monocytes behind to be collected from a layer as a peripheral blood mononuclear cell preparation [PBMC]) or differential lysis of red blood cells, leaving intact the more robust lymphocytes, monocytes, and neutrophils.

The light scatter signals (FSC and SSC) resulting from flow cytometric analysis of whole blood, lysed whole blood, and a mononuclear cell preparation after the density gradient separation of whole blood are compared in Figure 6.2. Each dot plot shows 2000 cells;

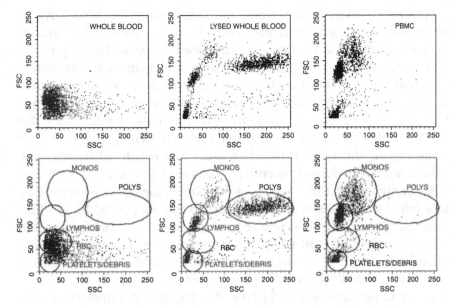

Fig. 6.2. The FSC and SSC signals resulting from the cells in different blood preparations (whole peripheral blood; whole peripheral blood after erythrocyte lysis; and peripheral blood mononuclear cells [PBMCs] with granulocytes removed by density gradient centrifugation). The bottom panels indicate the five clusters into which the scatter signals fall.

each cell is represented by a dot and plotted on the two axes according to the intensity of its FSC and SSC signals. The resulting clusters of particles consist of

1. A group of particles with low SSC intensity and low FSC intensity (and usually ignored because they fall below the useful FSC threshold)

2. A group of particles with low SSC intensity but somewhat higher FSC intensity

3. A group of particles, still with low SSC intensity but with moderately high FSC intensity

4. A group of particles with somewhat higher FSC but also with moderate SSC intensity

5. A group of cells possessing moderate FSC but much higher SSC.

The most obvious way to determine the identity of these particles, with distinctive scatter characteristics, is to sort them with a sorting flow cytometer (see Chapter 9). The flow sorting of blood preparations according to FSC and SSC parameters has brought the realization that the clusters of particles with defined flow cytometric scatter characteristics belong to the groups of cells as distinguished, traditionally, by microscopic anatomy. The anatomical differences that we see under the microscope result from different patterns of light bouncing off the cells on the slide and being registered by our eyes. Therefore, it should not be entirely surprising that a flow cytometer, with its photodetectors measuring light bouncing off cells, often clusters cells into groups that are familiar to microscopists. In addition, however, the human eye registers many more than two parameters when it looks at a cell (think of shape, texture, movement, color). Specifically, the eye registers something we might call *pattern* very sensitively; a flow cytometer registers pattern not at all. So, in some ways, it is perhaps just good luck that the two parameters of FSC and SSC light do allow cytometrists to distinguish some of the classes of white cells almost as effectively as a trained microscopist with a good microscope.

We should, however, always be aware of situations in which microscopists are better than cytometrists—certain types of classification are easy by eye but not at all easy by cytometer. For example, a microscopist would never confuse a dead lymphocyte with an erythrocyte, nor a chunk of debris with a viable cell; such mistakes are all too frequent in cytometry. Furthermore, microscopists, if they were not so polite, would find it laughable that cytometrists have a great deal of difficulty in distinguishing clumps of small cells from single large cells. However, if we think back to the discussion in Chapter 3 about the origin of the FSC signal, we can see why these problems occur and why we need to be aware of the limits of flow cytometry (and why a microscope is an essential piece of equipment in a flow cytometry lab).

The patterns of light scatter distribution illustrated in Figure 6.2 result from analysis of blood from a normal individual. With patterns like this, a flow cytometrist can, with some practice, set a so-called lymphocyte region around a group of particles that are mainly lymphocytes. This lymphocyte region will then, using FSC and SSC characteristics, define the cells that are to be selected (gated) for analysis of fluorescence staining. Such lymphocyte gating is the direct equivalent of a skilled microscopist's decision, based on size and nuclear shape, about which cells to include as lymphocytes in a count on a slide. A detailed discussion of gating is given at the end of this chapter. Because staining can provide us with some help in our gating decisions, we need first to discuss some of the staining techniques that can help us to describe cell populations.

STAINING FOR SURFACE MARKERS

Monoclonal antibody technology has provided flow cytometrists with a large array of antibodies that are specific for various proteins on the leukocyte surface membrane. The proteins (antigens), defined by these antibodies, have been given so-called CD numbers; "CD" stands for "cluster of differentiation" and refers to the group or cluster of antibodies all of which define a particular protein that differentiates cells of one type. More and more antigens are defined each year, and the CD numbers now range from CD1 on up through CD170 or greater. Antibodies against these antigens (called anti-CDx antibodies) have allowed immunologists to define taxonomic subgroups of leukocytes that are microscopically indistinguishable from each other but whose concentrations vary in ways that are related to an individual's immune status. For example, B lymphocytes and T lymphocytes look identical under the microscope and have similar FSC and SSC characteristics on the flow cytometer, but they possess membrane proteins that allow them to be distinguished by antibody staining (for example, B lymphocytes have CD20 on their surface; T lymphocytes have CD3). Some of the membrane antigens on leukocytes define function, some define lineage, some define developmental stage, and some define aspects of cell membrane structure whose significance is not yet understood. Although in this chapter I discuss the staining of leukocyte surface antigens with monoclonal antibodies, the principles apply equally well to the use of antibodies for staining any surface antigens on any type of cell.

An antibody will form a strong bond to its corresponding antigen.

To be of use in microscopy or flow cytometry, this bond needs to be "visualized" (to the eye or to the photodetector) by the addition of a fluorescent tag. Visualization can be accomplished by one of two different methods. With direct staining, cells are incubated with a monoclonal antibody that has been previously conjugated to a fluorochrome (for example, fluorescein or phycoerythrin or any fluorochrome with appropriate absorption and emission spectra). This procedure is quick and direct; it merely involves a half-hour incubation of cells with antibody (at 4°C), followed by several washes to remove weakly or nonspecifically bound antibodies. Cells thus treated are ready for flow analysis (although final fixation with 1% electron microscopic–grade formaldehyde will provide a measure of biological safety and long-term stability).

The second method (indirect staining) is more time consuming, less expensive, and either more or less adaptable depending on the application. Indirect staining involves incubation of cells with a non-fluorescent monoclonal antibody, washing to remove weakly bound antibody, and then a second incubation with a fluorescent antibody (the so-called second layer) that will react with the general class of monoclonal antibody used in the first layer. For example, if the primary monoclonal antibody happens to be an antibody that was raised in a mouse hybridoma line, it will have the general characteristics of mouse immunoglobulin; the second layer antibody can then be a fluoresceinated (fluorescein-conjugated) antibody that will react with any mouse immunoglobulin. The advantages of this two-layer technique are that, first, monoclonal (primary layer) antibodies are cheaper if they are unconjugated; second, a given second layer reagent can be used to visualize any monoclonal antibody of a given class (e.g., any mouse immunoglobulin, in our example here); third, comparison of antigen density according to fluorescence intensity is easier if common second layer reagents are used; and fourth, each step in the staining procedure results in amplification of the fluorescence intensity of the staining reaction. (With regard to this signal amplification that occurs with indirect staining, I might also add parenthetically that further amplification of weak signals is possible by using third and fourth layer reagents of appropriate specificity. All that is required is an interest in zoology. For example, if the primary monoclonal antibody is a mouse monoclonal, the second layer reagent could be a fluorescein-conjugated antibody raised in a goat and

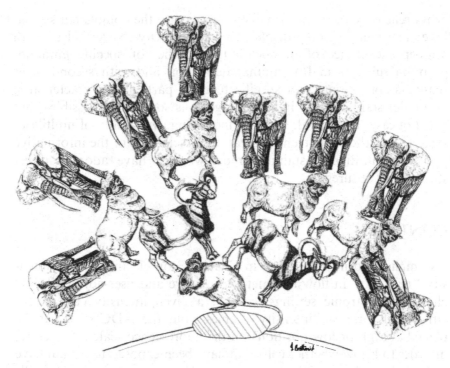

Fig. 6.3. Amplification of staining by the use of multiple antibody reagents. Drawing by Ian Brotherick.

specific for mouse Ig [known as a *goat anti-mouse reagent*]. An appropriate third layer reagent might then be a fluorescein-conjugated antibody raised in a sheep and specific for goat Ig [sheep anti-goat], and so on [elephant anti-sheep, armadillo anti-elephant, unicorn anti-armadillo] until the zoologists run out of immunologically competent animals [Fig. 6.3]. Because each antibody molecule is linked to many fluorochrome molecules and because many antibodies will bind to each antigen, this is a way to increase the intensity of signals from sparsely expressed membrane proteins.)

The disadvantages of indirect staining are that it is more time-consuming and it involves a second step that doubles the opportunity for nonspecific binding. Indirect staining also greatly limits the opportunity for simultaneous double and triple staining of cells with two and three different fluorochromes because of the problems of cross-reactivity between primary antigens and the conjugated second layer reagents that may display broad specificity. Nevertheless, with appropriate choice of monoclonal antibodies of specific immunoglobulin subclass and/or animal derivation and with second layer reagents appropriate and specific to these particular characteristics, two-color staining with indirect reagents is sometimes possible—but it is not easy. In general, with the increasing availability of multilaser systems and the concurrent realization by scientists of the informative power of multicolor staining, most workers have adopted direct staining procedures.

CONTROLS

As emphasized in the section in Chapter 3 on electronics, the intensity "read out" in flow cytometry is relative and user-adjustable. By changing electronic settings, cells of a given intensity can receive either high or low "intensity" values from the ADC (and can be placed at high or low positions on the fluorescence scale). Therefore, in order to know whether cells that have been exposed to a stain have actually bound any of that stain, we need to compare the stained cells with an unstained control. One of the general laws of science that applies particularly to flow cytometry is that no matter how many controls you have used in an experiment, when you come to analyze your results you always wish you had used one more. There are three reasons that this problem is acute in flow analysis. One has to do with the background fluorescence of unstained cells; the second has to do with the nature of antibody–antigen interactions; and the third has to do with the problem of compensation between overlapping fluorescence spectra from different fluorochromes.

The first problem that needs to be controlled is that of background fluorescence (called *autofluorescence*). All unstained cells give off some fluorescence (that is, all cells emit some light that gets through one or another of the filters in front of a cytometer's photodetectors). This autofluorescence may not be recognized by microscopists either because it is very dim or because experienced microscopists have acquired a mental threshold in the course of their training. But our cytometer's photodetectors are both very sensitive and completely untrainable. Therefore the autofluorescence of cells, resulting from intracellular constituents such as flavins and pyridine nucleotides, is bright enough to be detected. It can, in some cells, be so bright as to limit our ability to detect positive staining over and above this bright background. Whatever the level of this auto-fluorescence, we need to define it carefully by analyzing unstained cells (autofluorescence controls) if we are going to be able to conclude that cells treated with a reagent have actually become stained (that is, are now brighter than their endogenous background).

Beyond the problem of autofluorescence, there is a second problem. As discussed above, much of the staining of cells for flow analysis makes use of antibody–antigen specificity. Although the specificity between an antibody's binding site (the key) and the corresponding epitope on an antigen (the lock) is indeed exquisite, the beauty of the system can be confounded by a long floppy arm on the back (Fc) end of the antibody. These Fc ends stick with wild abandon to so-called Fc receptors that occur on the surface of many types of cells (notoriously monocytes). While I have worked in a department with scientists who study the specificity and importance of this Fc binding for too long to consider these reactions to be nonspecific and only a nuisance, it is true that these Fc receptors on many types of cells can confound the nominal specificity of an antibody's binding to its reciprocal antigen. What this means is that cells may stain with a particular monoclonal antibody because they possess a particular antigen on their surface membrane that locks neatly with the key on the monoclonal antibody binding site. They may also, however, stain with that particular monoclonal antibody because they possess Fc receptors that promiscuously cling to antibodies with all antigenic specificities. In addition, it is often true that dead cells (with perforated outer membranes) can soak up antibodies and then hang on to them tenaciously. The way to know if staining of cells is specific to a specific antigen is to use the correct control.

The correct control is always an antibody of exactly the same properties as the monoclonal antibody used in the experiment, but with an irrelevant specificity. If we are staining cells with a monoclonal antibody having a specificity for the CD3 protein occurring on the surface of T lymphocytes (and that monoclonal antibody happens to be a mouse immunoglobulin of the IgG_{2a} subclass, conjugated with six fluorescein molecules per molecule of protein and used to stain the cells at a concentration of 10 μg per ml), then an appropri-

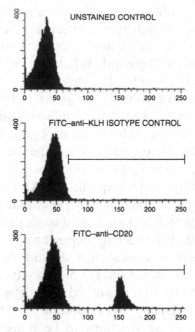

Fig. 6.4. The fluorescence histogram of an isotype control sample is used to decide on the fluorescence intensity that indicates positive staining.

ate control would be a mouse monoclonal antibody of the same sub-class, with the same fluorescein conjugation ratio, and at the same protein concentration, but with a specificity for something like key-hole limpet hemocyanin or anything else that is unlikely to be found on a human blood cell (Fig. 6.4).

Such a control antibody is known as an *isotype control* because it is of the same immunoglobulin isotype (subclass) as the staining anti-body used in the experiment. It will allow you to determine how much background stain is due to irrelevant stickiness (dead cells, Fc receptors, and so forth). The only trouble with this scenario is that exactly correct isotype controls are not usually available. Various manufacturers of monoclonal antibodies will sell so-called isotype controls and will certainly recommend that they be used. These are, however, general purpose isotype controls that will be of an average fluorochrome conjugation ratio and of a protein concentration that may or may not be similar to that used for most staining procedures. Whether an average isotype control is better than no isotype control at all is a matter of opinion and will depend on the kinds of answers that you demand from your experiments. For most routine immuno-phenotyping, where the staining of positive cells is strong and bright, isotype controls have been falling out of favor. Unstained (auto-fluorescence) controls may be good enough.

The third problem that needs to be controlled is that of spectral cross-over and the possibility of incorrect instrument compensation. As an example of a case in which controls for nonspecific staining, autofluorescence, and compensation are all critical, let us look at the staining of B lymphocytes for the CD5 marker present with only low density on their surface. As well as the problems created by non-specific staining and by autofluorescence, the problem of spectral cross-over between fluorescein and phycoerythrin can particularly confuse the interpretation of results from this kind of experiment. Look at Figure 6.5. What we are interested in is the number of B lymphocytes that possess the CD5 surface antigen. These cells will appear in quadrant 2 of a contour plot of fluorescein fluorescence

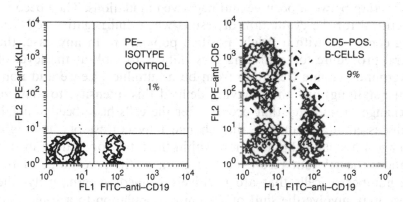

Fig. 6.5. The use of a phycoerythrin (PE) isotype control to help in deciding where, in a dual-color plot, to draw the horizontal line between fluorescein–stained cells to be considered positive and those to be considered negative for the PE stain. Mis-placing of the horizontal line will affect the number of CD19 cells determined to express the CD5 antigen in the stained sample. Data courtesy of Jane Calvert.

(a B-lymphocyte stain) on the horizontal axis against phycoerythrin (PE) fluorescence (the CD5 stain) on the vertical axis. But B cells will also appear in this quadrant if they have orange autofluorescence or if they are nonspecifically sticky for the anti-CD5 antibody (in this case a mouse monoclonal immunoglobulin of the IgG_{2a} isotype). In addition, they will appear in this quadrant if the cytometer's orange photodetector has not been properly compensated for cross-over from the fluorescein signal. The way around all these problems is to stain cells with a fluorescein stain for B cells in conjunction with an isotype control (a mouse IgG_{2a} antibody conjugated with PE but specific for an irrelevant antigen, say, keyhole limpet hemocyanin). The intensity of stain shown by these control cells on the PE photodetector will mark the limit of intensity expected from all nonspecific causes. Any further PE intensity shown by cells stained with the B-cell stain and the anti-CD5 PE stain will now clearly be the result of specific CD5 proteins on the cell surface. In this way, by use of the correct isotype control, we can rule out any problems in interpretation that may result from incorrect instrument compensation or nonspecific or background fluorescence.

In general, all these problems and their appropriate controls are particularly important when, as with the CD5 antigen on B cells, the staining density on the cells in question is low and there is considerable overlap between positive and negative populations. They become less critical for the evaluation of results when dull negative cells are being compared with a bright positive population. In any case, the general procedure for analyzing flow data is to look at the level of background staining (resulting from both autofluorescence and nonspecific staining) and then, having defined this intensity, to analyze the change in intensity that occurs after the cells have been stained. As discussed in Chapter 4, this change may consist of the bright staining of a small subpopulation within the total population; in this situation, the relevant result may be given as the percentage of the total number of cells that are positively stained. Alternatively, the change may involve the shift of the entire population to a somewhat brighter fluorescence intensity; here the relevant result may be expressed as the change in brightness (mode, mean, or median of the distribution). This leads us to the problem of quantification of intensity by flow cytometry.

QUANTITATION

One of the proclaimed advantages of flow cytometry, compared with eyeball microscopy, is its quantitative nature. Flow cytometry is indeed impressively quantitative when it comes to counting cells and compiling statistics about large numbers of cells in a short period of time. Users are, however, subjected to a rude shock when they first attempt to quantify the fluorescence *intensity* of their cells. Whereas a flow cytometer can be very quantitative about *comparing* the fluorescence intensity of particles (assuming that the photodetectors and amplifiers are working well), it is unfortunately true that a flow cytometer is very bad at providing an absolute value for the light intensity it measures. Therefore, any experimental protocol that needs to measure the intensity of the staining of cells (as opposed to a yes or no answer about whether and what percentage of cells are stained or not) is up against certain intrinsic difficulties.

If you really do need some measure of the intensity of cells, the way around these difficulties is to accept the limitations of the system, work within the constraints, and use some kind of standard to calibrate the intensity scale. The easiest standard for any cell is its own unstained control. An arbitrary position on the scale can be assigned to the fluorescence of the control (by changing the voltage on the photomultiplier tube during instrument set-up), and the stained sample can be compared with this. The disadvantage in this measure of relative fluorescence compared with the control is that cells with high autofluorescence will require a greater density of positively stained receptors to give the same "relative intensity" value as cells with low autofluorescence. In other words, if intensity is expressed by a ratio of the brightness of stained cells relative to that of the unstained cells, a given ratio will represent more positive stain (in terms of fluorochrome molecules) on highly autofluorescent cells than on cells with low background.

One way around this problem is to compare cell fluorescence not with unstained cells but with the fluorescence of an external standard. This can be done by the use of fluorescent beads. In brief (this is an insiders' joke; you would be amazed at how much has been written about the use of beads in flow cytometry), there are commercially available polystyrene beads ("microspheres") that have standardized fluorescence intensities. Some of these beads have fluorescein or PE bound to their surface. Others have a selection of hydrophobic fluorochromes incorporated throughout the bead. The latter are more stable in intensity, but, because the fluorochromes are not the usual flow cytometric fluorochromes, they may give different relative values on the different photodetectors of different cytometers. In either case, by running a sample of standard beads through the cytometer, all data can be reported as a value relative to the intensity of the standard beads.

One further step toward calibration has been taken with the use of a calibration curve made with sets of beads with known numbers of fluorochromes on their surface. Such calibrated beads are available with known numbers of PE molecules. Similar, but less direct, beads are available with fluorochrome molecules that have been calibrated in units equivalent to the intensity of fluorochrome molecules in solution ("MESF" units = "molecular equivalents of soluble fluorochrome"). With these beads, a curve can be obtained (Fig. 6.6), giving each channel on the ADC a calibration in number of fluorochrome molecules (for PE) or MESF values (for fluorescein). In this way, the background fluorescence of a control sample can be expressed as an equivalent number of fluorochrome (or MESF) molecules and can be subtracted from the number of fluorochrome molecules of a stained sample. The fluorescence of the stained sample can then be expressed as, for example, PE molecules over and above the background level.

Having now determined a value that might, with luck, quantify the brightness of a particle in terms of fluorochrome molecules or soluble equivalents, one may wonder how best to convert that value into the number of receptors or antigens on the surface of the cell. At first thought, calculation of this value might be determined if values are known for the number of fluorochrome molecules per antibody (the F/P ratio) used in the staining procedure. Unfortunately, even if this value has been determined chemically, it will not apply within a system in which there is quenching of the fluorescence from fluorochromes in closely packed regions on a cell surface (causing a bound fluorochrome to fluoresce considerably less brightly than in its soluble form). Moreover, the F/P value will almost certainly not be known in a system with indirect staining and undetermined amplification. At the present time, this F/P value can only be used with confidence in certain staining systems where antibodies have been certified to contain a single PE molecule per antibody and are used in

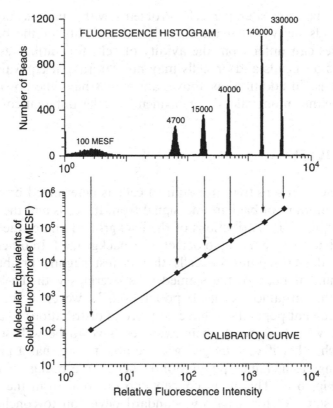

Fig. 6.6. The fluorescence histogram of a mixture of fluorochrome-conjugated calibration beads and the calibration line for channel numbers and their equivalence in soluble fluorescein molecules derived from that histogram. From Givan (2001).

conjunction with standard beads with known numbers of PE molecules. Even in this case, the final value will be in terms of the number of antibodies bound to a cell, and this may not be easily related to the number of receptors per cell (because antibody binding to receptors may be monovalent or bivalent).

Other types of calibration help may be available in the form of a different type of calibrated microsphere. Beads can be obtained that possess a known number of binding sites for immunoglobulin molecules. They can be treated as if they were cells and stained in the routine way with the antibody stain (direct or indirect) in question. The intensity of the beads with known numbers of antibody binding sites can be used to calibrate the scale, converting ADC channels to antibody binding sites per cell. Problems with quantification using these beads derive from the fact that the avidity of the beads for antibodies can differ from the avidity of cells for antibodies so that receptors on beads and on cells may not saturate at equivalent concentrations. In addition, as above, antibodies have the possibility of binding either monovalently or bivalently under different conditions.

SENSITIVITY

Light detection sensitivity for stained cells is determined by two factors: the amount of background signal from the cells and the breadth of the population distributions of the background and of the positive signals that you are trying to detect over background. In other words, you can detect staining on cells that is just slightly brighter than background if none of the stained cells overlaps with the brightest cells in the unstained (control) population. However, if the control and fluorescent populations have very broad distributions (that is, the range of values is great), their averages have to be well separated from each other if you are going to be able to say that a particular cell belongs to a stained population rather than a background population (Fig. 6.7). This is, in concept, no different from the requirement in statistics for a narrow standard deviation to conclude that two populations with closely similar means are significantly different from each other, but a less stringent requirement for narrow standard deviation if the two population means are well separated.

As a practical matter, the way to make a flow cytometer more sensitive in detecting weak light signals is to lower the noise in the background by using good optical, fluidic, and electronic components and to align the instrument well so that fluorescence detection efficiency is high and fluorescence distributions are as narrow as possible. The result of these considerations leads, however, to the unavoidable conclusion that cells with intrinsically high background make it more difficult to detect low numbers of fluorochromes derived from the staining procedure. Think of trying to see the stars in the daytime. Certain classes of cells have greater autofluorescence than others as a result of their metabolic activity. All other things being equal, large cells have more autofluorescence than small cells, simply because they are larger and have more autofluorescent molecules associated with each cell.

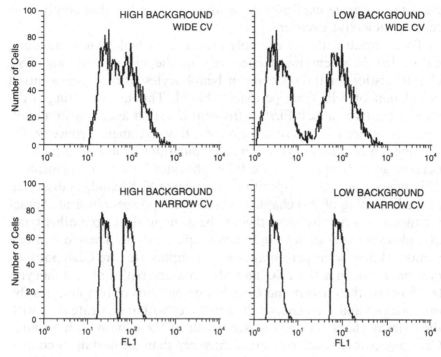

Fig. 6.7. The ability to distinguish stained cells from unstained cells depends on both the breadth of the distributions of light intensities of the two populations as well as their relative average intensities.

THE STRATEGY OF GATING

In flow cytometry the term *gating* is applied to the selection of cells (according to their fluorescence and/or scatter characteristics) that will be carried forward for further analysis. The process of gating corresponds to the decisions made by the microscopist about what particles in a field to include in a count of any particular type of cell. Gating is generally acknowledged to be one of the most powerful, but also one of the most problematic, aspects of flow cytometry (right up there with spectral compensation). It is problematic primarily because flow cytometrists like to think of their technique as objective and do not like to admit that much of flow cytometric analysis rests on the foundation of a few subjective initial decisions. The ideal strategy for gating should therefore move us toward two goals: First, gating needs to become as objective as possible, and second, flow cytometrists need to recognize explicitly those aspects of gating that continue to require subjective decisions.

To continue with the example drawn from leukocyte techniques, gating has been employed effectively in the use of FSC and SSC characteristics for the selection of lymphocytes from within a mixed population of cells from peripheral blood. This use of gating is derived from two causes: First, a frequent question asked by immunologists (whether of a microscope or of a flow cytometer) concerns the distribution of lymphocytes into subpopulations. For example, "what percentage of lymphocytes are B lymphocytes?" or "what percentage of lymphocytes are CD8-positive lymphocytes?" Second, as discussed at the beginning of this chapter, lymphocytes possess physical characteristics that generally allow them to be distinguished from other types of leukocytes in both flow and microscopic analysis. Therefore, if one wants to know what percentage of the lymphocytes are CD8 positive, one simply defines the FSC and SSC characteristics of lymphocytes (by flow) or the nuclear and cytoplasmic patterns of lymphocytes (by microscope) and then analyzes the particles within this gate (i.e., with the defined characteristics) to see which of these stain with an anti-CD8 monoclonal antibody more intensely than an unstained control (Fig. 6.8). The ability to define a lymphocyte gate takes a bit of practice, but it is usually an easy decision for a skilled flow cytometrist (and also for a skilled cytologist using a microscope) when

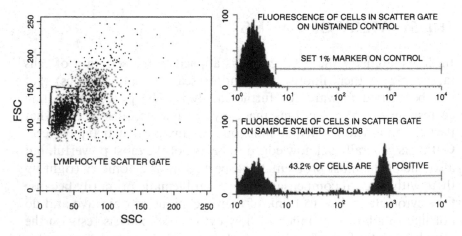

Fig. 6.8. A common type of flow analysis. The cells within a lymphocyte gate are analyzed, first in an unstained control sample and next in a stained sample, to determine how many cells within that gate are positively stained.

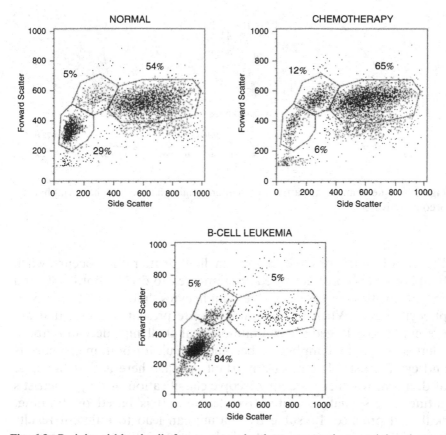

Fig. 6.9. Peripheral blood cells from a normal volunteer, a patient receiving chemo-therapy, and a patient with B-cell leukemia. Normal cells appear as a tight cluster of lymphocytes with a diffuse group of monocytes on their shoulder possessing brighter FSC and SSC and a large cluster of cells with high SSC (neutrophils). The patient receiving chemotherapy has clusters in similar positions, but with far fewer lympho-cytes that merge at their top end into the monocytes. The leukemic patient's cells are almost exclusively lymphocytes, which have slightly lower FSC than normal cells. Data files were provided by Marc Langweiler and Sharon Rich.

blood is relatively normal. It can, however, be a very difficult deci-sion, by either technique, when blood is abnormal (e.g., blood from immunosuppressed patients, who have relatively few and possibly ac-tivated lymphocytes).

Figure 6.9 shows the light scatter profiles from cell preparations from a normal donor and from patients with leukemia and receiving chemotherapy, just to show diverse examples of light scatter profiles.

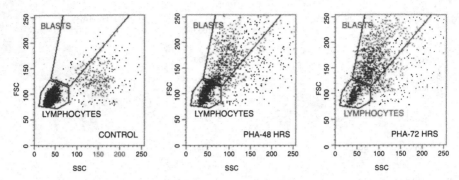

Fig. 6.10. The changes in patterns of scattered light that occur when lymphocytes become activated.

Figure 6.10 indicates the change in light scatter that occurs when lymphocytes become activated or stimulated to divide; both FSC and SSC intensities increase, and the lymphocytes "move out of the lymphocyte gate." When few lymphocytes are present or those that are present have enlarged after immunological challenge, decisions about what is or is not a lymphocyte become difficult for both microscopists and cytometrists. The most important message here is that this type of decision, whether it be microscopic classification or a cytometrist's gating, is essentially a subjective decision. It is based on training, skill, and practice. Its subjective nature can lead to different results from different operators, particularly when operators are inexperienced or when abnormal specimens are being analyzed, because the percentage of cells within a gate that are stained (with any given stain) will vary depending on the range of cells included within that gate (see Table 6.2). If we want to analyze lymphocytes, then the gate should ideally include all the lymphocytes from the sample and should include only lymphocytes.

Recognition of the need for some objectivity in gating decisions (both because operators may not be experienced and because not all samples are "normal") has led to attempts both to automate the defining of the gate and to describe the goodness of the gate, once defined. Progress toward automation with lymphocytes brings us to the technique of back-gating, which has proved informative as a strategy for flow cytometric analysis in general. Historically, gating has been performed on the SSC and FSC characteristics of cells (by

TABLE 6.2. The Effect of Size of the FSC/SSC Gate on Determination of the Characteristics of Cells Within that Gate: Peripheral Blood Mononuclear Cells from a Heart/Lung Transplant Patient

	Small gate (691 cells)	Large gate (1074 cells)
Lymphocytes	646 cells	928 cells
	94% of gated cells	86% of gated cells
CD3 cells (T cells)	81% of gated cells	70% of gated cells
IL-2r$^+$ (activated) T cells	12% of T cells	16% of T cells
CD20 cells (B cells)	9% of gated cells	8% of gated cells
CD4 cells (helper T cells)	56% of gated cells	51% of gated cells
CD8 cells (cytotoxic T cells)	24% of gated cells	20% of gated cells
CD16 (NK cells)	8% of gated cells	11% of gated cells

analogy with a microscopist's use of physical criteria to classify cells), and this gate has then been used to inquire about the fluorescence properties of the gated cells. Back-gating is really just a different sort of gating—it reverses the usual protocol by using a stained sample, placing a gate around particles with certain fluorescence character-istics, and then asking what the physical characteristics (i.e., FSC and SSC) of these particles are. Once the scatter characteristics of a population have been ascertained in this way, the subsequent placing of the scatter gate can be done on the basis of rational and defined criteria.

For example, if one stains leukocytes with a PE-conjugated monoclonal antibody specific for the CD14 determinant (a monocyte marker) and puts a gate around the cells that are brightly fluorescent, it can be determined where these bright cells fall in a plot of FSC versus SSC (Fig. 6.11). As it turns out, most of the CD14-positive cells have moderate SSC and moderate FSC and lie in a cluster on the FSC/SSC dot plot. It also turns out, however, that there are a few CD14-positive cells that lie outside this region. We are now in a position to count the total number of CD14-positive cells in the sample; to draw what we think is an appropriate gate within the FSC/SSC plot; and finally to ask two questions that will tell us how good a gate we have drawn: (1) How many of the CD14-positive cells have we excluded by the drawing of that gate? (2) What percentage of the cells within the gate are not monocytes? Thus the technique of back-gating has allowed us to make an "educated guess" in drawing

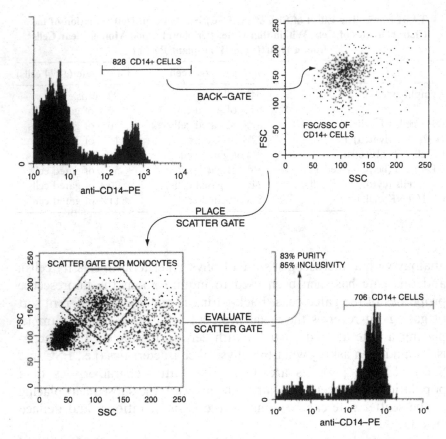

Fig. 6.11. Back-gating from CD14 fluorescence to determine the scatter characteristics from monocytes. Such back-gating facilitates the placing and then evaluation of a monocyte scatter gate.

a monocyte gate in the FSC/SSC plot; and it has then allowed us to evaluate that gate in terms of both purity and inclusivity. A perfect monocyte gate would contain only monocytes (100% purity) and would include all the monocytes (100% inclusivity). As we shall see, most gates are a compromise between these two goals. In any case, with back-gating to help us define a monocyte (FSC/SSC) gate, that scatter gate can be used in subsequent samples when staining the monocytes to analyze their phenotype.

By extension from this simple example, the technique of back-gating has been applied quite elegantly to leukocytes by the use of a combination of two stains and two-color analysis. When a PE-conjugated monoclonal antibody specific for monocytes (as described above) is mixed with a fluorescein isothiocyanate (FITC)-conjugated antibody specific for a determinant found on all white cells (CD45 is a so-called leukocyte common antigen), this mixture can be used to stain a white cell preparation. After dividing the FITC/PE plot into four quadrants, it can be imagined that any erythrocytes or debris in the preparation will appear in the lower left quadrant (double nega-tives); any white cells will appear in the upper or lower right quad-rants (FITC positive); and any monocytes will appear in the upper right quadrant (double positive). One might imagine that lympho-cytes and granulocytes would appear in the lower right quadrant, as they are leukocytes but not monocytes (and should be FITC positive, PE negative).

The results from such a staining protocol are given in Figure 6.12. It turns out that they are even more useful than we might have imagined. It happens that granulocytes express a lower density of the common leukocyte antigen on their surface than do lymphocytes, and these two types of cells can be distinguished from each other as well as from the PE-positive monocytes in this staining mixture. Thus this protocol allows us to do the two things we have set as goals for our lymphocyte gating strategy. We can back-gate from the lymphocyte cluster in the FITC/PE plot to help us find the FSC/SSC region in which to draw our gate, and we can use the staining profile to eval-uate that gate in terms of purity and inclusivity. In the case of the example shown in Figure 6.12, back-gating from the FITC-bright/PE-negative cluster leads us to a region for the FSC/SSC gating of lymphocytes. If we draw a reasonable scatter gate based on the scatter signals of the lymphocytes, we find that we have included most of the lymphocytes and that almost everything in that gate is a lymphocyte. A smaller gate might have higher purity but miss some lymphocytes; a larger gate might include 100% of the lymphocytes but some mono-cytes, neutrophils, and red blood cells as well. This kind of staining and back-gating protocol has led the way toward a fully automated gating procedure in which the gating and evaluation of that gate are done by computer. A computational gating algorithm can be written to adjust the scatter gate to aim for some desired level of purity and/or inclusivity.

However, any automated procedure will have some degree of

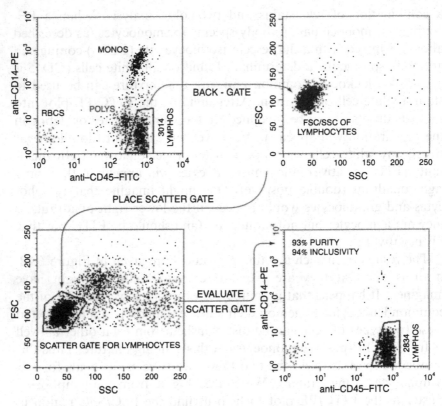

Fig. 6.12. Back-gating from CD14/CD45 fluorescence to determine the scatter characteristics of lymphocytes. Such back-gating facilitates the placing and then evaluation of a lymphocyte scatter gate within a peripheral blood mononuclear cell preparation.

difficulty with abnormal samples, and laboratories will tend to have their own (subjective) methods for compromising when the sample is such that a gate with both high purity and high inclusivity is not possible. By way of illustration of this situation, we can look at a sample of blood cells from an immunosuppressed patient (Fig. 6.13). Back-gating from the FITC-bright/PE-negative cluster leads us to placing a gate in a region of the FSC/SSC plot. We can see, however, that there appear to be lots of lymphocytes that have abnormally high FSC and SSC; these could be activated lymphocytes. If we draw a gate large enough to include these lymphocytes in our subsequent

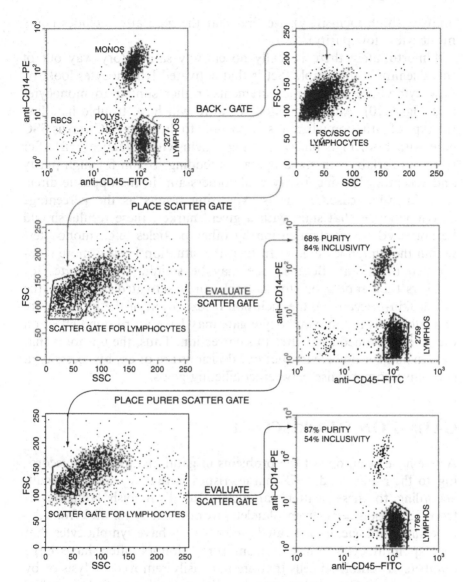

Fig. 6.13. Difficulties in placing a pure and inclusive lymphocyte scatter gate on a peripheral blood mononuclear cell preparation from a transplant patient with few and blasted lymphocytes.

analysis (high inclusivity), we find that the gate also includes many monocytes (low purity).

Unfortunately there is really no entirely satisfactory way out of this dilemma—the simple fact is that activated lymphocytes look, by flow cytometric scatter measurements, rather similar to monocytes (see Fig. 6.10). Any automated software will have trouble handling the type of situation when it is impossible to draw an FSC versus SSC gate with both high purity and high inclusivity. Practice will differ from lab to lab, with some operators tending to aim for high purity and accepting low inclusivity while others aim in the opposite direction. In either case, if one is expressing results as the percentage of lymphocytes that stain with a given marker, these results should be corrected for contamination by other particles (e.g., monocytes) within the lymphocyte gate. In fact, the situation is even more complicated than that. Because one may be staining lymphocytes for markers that appear on, for example, only activated cells (e.g., the interleukin-2 receptor), the inclusion or exclusion of the larger, more activated cells in the lymphocyte gate may have a profound effect on the result obtained even after this correction. Thus, the upshot is that we can aim for objectivity, but our decisions are often, by necessity, a subjective compromise between conflicting goals.

GATING ON FLUORESCENCE

As we have seen, one of the problems in gating lymphocytes according to their FSC and SSC characteristics is that it can be difficult, according to these scatter parameters, to distinguish lymphocytes from red blood cells, from platelets, from monocytes, and from debris. Immunologically activated patients may have lymphocytes that overlap monocytes; samples from some donors often have large numbers of red blood cells that are not easily removed by lysis or by centrifugation. Gating these preparations according to FSC and SSC may result in fluorescence analyses being reported as the stained percent, not of lymphocytes, but of an unknown mixture of cells.

For example, the more red blood cells that overlap lymphocytes in the scatter gate, the lower will be the percentage of those gated cells that are scored as CD4-positive (even though the percentage of lymphocytes that are CD4-positive may be identical in all cases). Now

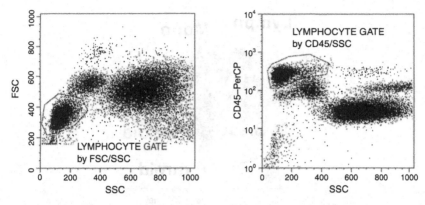

Fig. 6.14. The use of CD45 to exclude erythrocytes, platelets, and debris from a lymphocyte gate. The distinction between monocytes and lymphocytes can be ambiguous with either the FSC/SSC or the CD45/SSC gate. Data file provided by Marc Langweiler.

that many benchtop flow cytometers have "parameters to spare," current clinical protocols for lymphocyte analysis suggest that antibodies against CD45 be added to all tubes as a third color to be used for gating. By gating on CD45 positivity along with SSC, the gated cells (Fig. 6.14) will not include high numbers of erythrocytes, debris, or platelets (all CD45 negative). In this way gating can be refined if you are using a flow cytometer with extra fluorescence parameters. The distinction between lymphocytes and monocytes is not, however, helped by this procedure (and the CD45/SSC gate in Fig. 6.14 indicates ambiguity here). Figure 6.15 illustrates CD45/SSC gating in a bone marrow sample, where this type of analysis has proved particularly useful in identifying clusters that are related to various mature and immature hematological lineages.

In fact, this CD45/SSC gating marks the general trend away from using scatter parameters toward a quite different strategy for flow analysis. The basic problem, as seen above, is that lymphocytes (or any other taxonomic group) are not a homogeneous collection of cells with perfectly delineated physical characteristics; they are mainly homogeneous, but they usually contain at least some cells at the fringes with marginal characteristics. Any gating based on FSC/SSC forces us either to exclude these fringe cells or to include many extraneous cells. With the availability of multicolor instrumentation

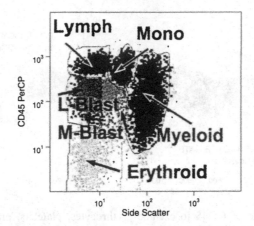

Fig. 6.15. Bone marrow from a normal donor showing CD45/SSC clusters. CD45 expression varies as cells mature. Lymph = lymphocytes; Mono = monocytes; L-Blast = lymphoblasts; M-Blast = myeloblasts; Myeloid = neutrophils; and Erythroid = cells of the erythroid lineage. From Loken and Wells (2000).

and multicolor stains, it has become possible to avoid making any of these difficult gating decisions and to include all cells in the analysis by using stain itself either to gate in or to gate out the cells of interest.

For example, if one is studying the prevalence of a particular subpopulation among lymphocytes, one could stain every peripheral blood mononuclear cell sample with a PE stain for monocytes in addition to an FITC conjugate of the marker of interest. Then, in the analysis stage, one could simply gate out (exclude) from analysis any PE-positive particles and analyze all the PE-negative particles for the percentage that are FITC-positive. Figure 6.16 shows an example of this protocol. By the use of three-color analysis, there are even greater possibilities. This type of analysis avoids the necessity of prior decisions about the FSC and SSC characteristics of the cells of choice (e.g., lymphocytes). It becomes particularly important when the cells of interest are less homogeneous in physical characteristics than lymphocytes. For example, by gating on a stain that is specific for cytokeratin (a protein found on tumor cells of epithelial origin), tumor cells within a mixed population from a breast tumor biopsy specimen can be selected for further analysis (e.g., DNA content) without any prejudgement about the FSC or SSC of a poorly defined and heterogeneous population of abnormal cells.

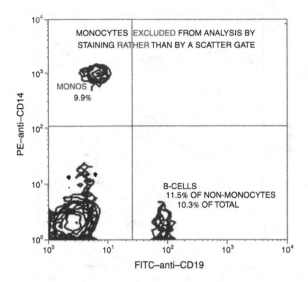

Fig. 6.16. Rather than a scatter gate, a PE-anti-CD14 stain can be used to exclude monocytes from a count of B and non-B lymphocytes. Data courtesy of Jane Calvert.

A logical extension of this kind of technique can be seen in the methodology proposed by PK Horan in 1986: No decisions based on the scatter characteristics of cells have been made. Cells are stained simply with a cocktail of conjugated monoclonal antibodies at appropriate (sometimes not saturating) concentrations, and all the cells in the mixture are classified according to their staining characteristics (Fig. 6.17). Staining cells from leukemic patients currently follows a similar strategy—where abnormal and normal cells are defined by the way they cluster in a two-dimensional dot plot. In other words, cells are defined by their relative intensities more than by their negativity or positivity for a given antibody.

At the beginning of this section, a strategy was described for placing a scatter (FSC/SSC) gate and then evaluating it in terms of purity and inclusivity. We were then forced to admit that gating is often an uneasy and subjective compromise between these conflicting criteria. We find therefore that we must conclude that using FSC and SSC characteristics to gate cells may not be a good thing after all. The availability of multicolor analysis has led to a trend toward using staining characteristics to define the cells of interest without regard to

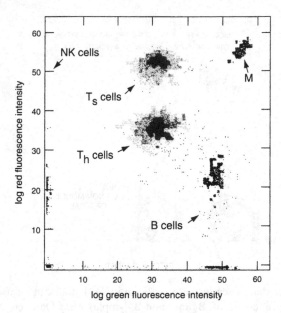

Fig. 6.17. The two-color fluorescence profile of peripheral blood mononuclear cells stained simultaneously with six different monoclonal antibodies to delineate five different populations of cells. From Horan et al. (1986).

their possibly variable physical (light scatter) properties. By using one or more colors in the analysis either to select or to exclude (gate in or gate out) particular groups of cells, we can avoid any prejudgement on the physical characteristics of those cells. This overall strategy makes sense when our system allows for multicolor analysis (with parameters to spare) and when antibodies are available to define the cells of interest. It is, nevertheless, still true that antibodies to define subsets of cells are not perfect: Cells of different phenotypes react with similar antibodies (sometimes, thankfully, at different intensities), and several antibodies are often required to fully define the taxonomy of a cell.

As a summary comment on gating, we need simply to remember that in flow cytometry our questions are usually formulated in terms of "what percentage of a certain population of cells is positive for a certain set of characteristics?" The choice of a gate defines that "certain population" of cells. The choice of that gate will therefore affect the answer to the question (the percentage positive). Whether gating is applied by means of scatter characteristics, by means of staining characteristics, or not at all, the procedure still needs to be described and quantified if the results are to be meaningful and reproducible. It is only when we have stated exactly which "certain population of cells" we are analyzing that we have fulfilled our goals of objectivity and/or *explicit* subjectivity in flow analysis.

FURTHER READING

Chapters 3 and 5 in **Ormerod**, Chapters 3.2 and 3.3 in **Diamond and DeMaggio**, Chapters 10 and 11 in **Darzynkiewicz**, Chapter 6.2 in **Current Protocols in Cytometry**, and Chapters 17 and 34 in **Melamed et al.** are all good discussions of general lymphocyte staining methodology for flow analysis.

Chapter 1.3 in **Current Protocols in Cytometry** and Chapter 14 in **Darzynkiewicz** (both by Robert Hoffman) are excellent discussions of sensitivity and calibration.

Volume 33, number 2 (1998) of **Cytometry** is a special issue devoted to "Quantitative Fluorescence Cytometry: An Emerging Consensus."

7

Cells from Within: Intracellular Proteins

Although many applications of flow cytometry involve the staining of cells for proteins expressed on the outer membrane, cells also have many proteins that are not displayed on their surface. With appropriate procedures, flow cytometry can provide a means to analyze these intracellular proteins. The outer cell membrane is impermeable to large molcules like antibodies; however, if we intentionally fix cells to stabilize proteins and then disrupt the outer membrane, the cells can be stained with fluorochrome-conjugated monoclonal antibodies against intracellular proteins. After time to allow the antibodies to pass through the now-permeabilized membrane, the cells are washed to remove loosely bound antibodies and then are run through the flow cytometer to measure their fluorescence intensity.

This intensity should, under good conditions, be related to the amount of the intracellular protein present. However, in describing our ability to stain cells for surface proteins, we mentioned that it is best to stain viable cells. Dead cells have leaky outer membranes; they often show high nonspecific staining because antibodies get through the disrupted membrane and become trapped in the intracellular spaces. Therein lies a conflict in our ability to stain cells for intracellular proteins. Because antibodies of all types are easily trapped in the cytoplasm, there is greater potential for nonspecific staining of permeabilized cells than intact cells. The very procedure that we carry out to give access of the staining antibody to its target (intracellular) antigen actually increases the access of all antibodies to nonspecific targets. To lower this nonspecific background, antibody titers are critical and washing steps are important. Unfortunately, even with low antibody concentrations and careful washing, background fluorescence from isotype-control antibodies is often considerably higher on permeabilized than on intact cells.

There is, in addition, a second problem. The procedures used for fixing and permeabilizing cells—to give the staining antibodies access to intracellular proteins—can modify or solubilize some antigens, thus destroying the stainability of the very proteins that are being assayed. To make matters worse, the protocol that works best for one antigen may entirely destroy a different antigen. This should not be surprising after consideration that "intracellular" includes proteins of many types and in many different environments. Some intracellular proteins are soluble, some are bound to organelle membranes, and some are in the nucleus. Therefore, methods for staining cells for intracellular proteins cannot be as standard or as dependable as the methods for staining surface proteins. They have to be individually optimized for the cells and the proteins in question.

METHODS FOR PERMEABILIZING CELLS

While not attempting to describe possible methods in detail, I feel it is important here to point out the issues involved in intracellular staining because they highlight some general issues that affect all of flow cytometric analysis. Methods for permeabilizing and fixing cells are various and must be optimized for the particular intracellular antigens being detected because some antigens are more robust than others in the face of different agents. Figure 7.1 gives an example comparing fixation/permeabilization effects on two different intracellular antigens: Five different fixation/permeabilization protocols have been used, and their effects on staining PCNA (proliferating cell nuclear antigen) and p105 (a mitosis-associated protein) have been compared. The good news is that you can stain for intracellular antigens. The bad news is that it may be difficult to stain cells optimally for two different antigens at the same time (and the relative intensity of staining for two different antigens may tell you little about the actual relative proportions of these proteins in the cell).

The general protocol for intracellular staining involves, first, staining the cells for any surface (outer membrane) antigens, as described in the previous chapter. Then the surface proteins with their

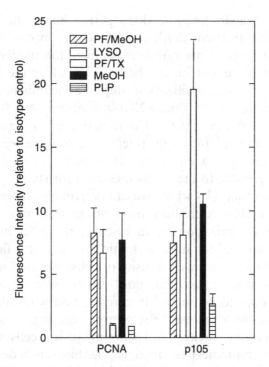

Fig. 7.1. The effects of different fixation protocols on the relative amounts detected of two different intracellular proteins. Modified from A McNally and KD Bauer as published in Bauer and Jacobberger (1994).

bound antibodies, as well as the intracellular proteins, are fixed gently to stabilize them. The purpose of the fixation is to cross-link the proteins well enough that they are not removed or washed out of the cells after the cells are permeabilized, but not so well that the intracellular antibody binding sites are masked or destroyed. Although ethanol and methanol can be used for fixation (by themselves or following another fixative), the most common fixative used prior to intracellular staining is formaldehyde. Formaldehyde is generally used at lower concentration and/or for a shorter period of time than for routine fixation of surface-stained cells (where fixation overnight in 1% formaldehyde is the [optional] last step of the procedure before flow cytometric analysis). Formaldehyde (at 0.5–1.0%) for 10 min is a good suggested concentration and time for cell fixation, but lower or higher concentrations, for shorter or longer periods of time, might be required.

This formaldehyde fixation does permeabilize the cytoplasmic membrane a bit (formaldehyde-fixed cells are permeable to small molecules), but proteins are often cross-linked too tightly for staining of intracellular proteins with antibodies. Therefore the fixation step is followed by a permeabilization step. Permeabilizing agents are usually detergents, such as Triton X-100, digitonin, NP40, or saponin, at concentrations of about 0.1%. Combined fixation/permeabilization reagents are also available as proprietary commercial reagents. With luck, the detergent will open up the cell enough so that the now-fixed proteins are accessible to the antibodies used for staining.

What are the criteria by which we can determine whether a fixation/permeabilization procedure has been optimized for an antigen in question? This optimization is, in essence, no different from optimization of a protocol for surface staining of cells. It is first necessary to maximize the fluorescence intensity of cells that are known to possess the intracellular antigen (the positive control); fixation time and concentration need to be altered in combination with different detergent concentrations to increase the positive staining. It is then necessary to decrease the background staining (using cells stained with isotype-control antibodies) as much as possible; this is done by trying increasing detergent concentrations and washing the cells thoroughly in buffer that contains the detergent. In other words, the goal is to increase the signal-to-noise ratio. Because antibody concentration, fixative agent, fixative concentration, fixation time, choice of permeabilization agent, and concentration of that permeabilizing agent are all variables in this protocol (and the optimal characteristics of each may be different for different antigens), staining for intracellular antigens requires some persistence on the part of the investigator. The following examples (in this chapter and in the following chapter on DNA) will demonstrate, however, that it is certainly possible.

EXAMPLES OF INTRACELLULAR STAINING

From the point of view of a flow cytometer, surface, cytoplasmic, and nuclear proteins are similar. The flow cytometer cannot ascertain the location of the source of fluorescence. In addition, the nuclear membrane has large enough pores that it provides little or no obstacle to staining once the outer, cytoplasmic membrane has been breached.

Cells have been stained successfully for nuclear proteins related to proliferation (for example, PCNA, Ki-67, and various cyclins, which will be discussed in the chapter on DNA) and to tumor suppression (for example, p53, c-myc, and the retinoblastoma gene product). They have also been stained for proteins bound to interior membrane surfaces (e.g., Bcl-2, multidrug resistance protein [MDR], and P-glycoprotein), and many strictly cytosolic proteins have been analyzed (like tubulin, hemoglobin, surface proteins that exist intracellularly at various stages of differentiation, and many cytokines).

As an example of one of the more complex biological situations, we can use the staining of cytokines as an illustration. Cytokines are a diverse class of proteins that, in response to cell stimulation, are synthesized and then secreted by leukocytes. For example, when T lymphocytes are stimulated, either nonspecifically or by immunological triggers, they begin to synthesize interferon-γ in their endoplasmic reticulum, send the proteins to the Golgi apparatus, and then secrete the molecules into the environment for stimulation of neighboring cells. To stain for intracellular interferon-γ, the usual technique is to stimulate cells with a biological trigger and then to incubate them with an inhibitor (brefeldin A or monensin) for several hours. These inhibitors block the normal secretion of proteins from the Golgi apparatus and thus allow the cytokine concentration to build up in the cell to levels that are detectable. After the incubation period, the cells are stained for any surface antigens of interest, fixed briefly in formaldehyde, permeabilized with saponin, and, finally, stained with a monoclonal antibody against interferon-γ.

Figure 7.2 shows an example of the way in which cells can be stained for a phenotypic surface marker (CD8) as well as the intracellular cytokine, interferon-γ. The flow data indicate that interferon-γ is associated, after PMA-ionomycin stimulation, primarily with CD8-negative cells. More of the CD8-negative than the CD8-positive cells have intracellular interferon-γ, and those that have that cytokine have more of it per cell. The tricks in the procedure for staining intracellular cytokines are as much biological as chemical (because the stain is for the end result of a functional process). In addition to a knowledge of how to fix and permeabilize a cell and how to avoid nonspecific staining, we require knowledge of how to trigger the cytokine production, knowledge of the time course of cytokine synthesis after stimulation, and knowledge of how long cells can survive

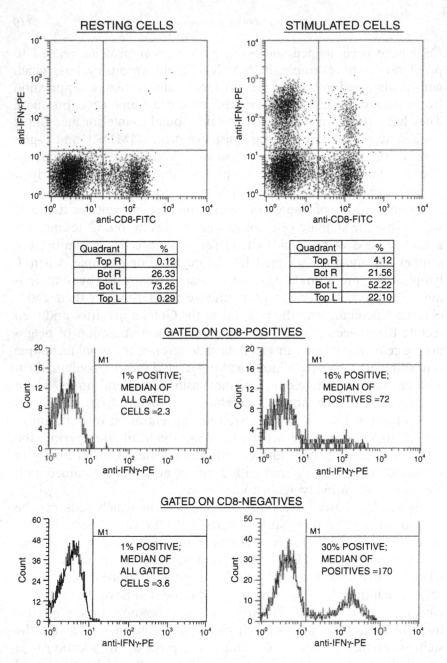

Fig. 7.2. Dot plots showing the staining of lymphocytes for intracellular interferon-γ in conjunction with an outer membrane stain (against CD8) to phenotype the cytokine-producing cells. Cells were stained for CD8 and then fixed with formaldehyde and permeabilized with saponin. The stimulus was PMA-ionomycin. Data courtesy of Paul Wallace.

with brefeldin A or monensin inhibition so that they build up large amounts of easily detectable cytokines but do not burst from this dire treatment.

As another example of intracellular staining, we can look at data from the staining of human breast tumor cells for cytokeratin and for the estrogen receptor (both intracellular proteins). Tumor cells were obtained following mastectomy by mincing and sieving the tissue to form a single-cell suspension. The suspension was then treated with saponin to permeabilize the cells. After staining for both cytokeratin and the estrogen receptor, cytokeratin-positivity selects the cells in the mixture that are of epithelial tumor origin (excluding stromal or infiltrating inflammatory cells). The two-color plot in Figure 7.3 indicates that the cytokeratin-positive (but not the cytokeratin-negative) cells express the estrogen receptor strongly (estrogen receptor positivity is associated with superior prognosis and a greater responsiveness to endocrine therapy). Gating on the cytokeratin-positive cells permits the analysis of tumor cells by themselves for the estrogen receptor without concern about the variable contamination of tumor cells by stromal cells in different samples.

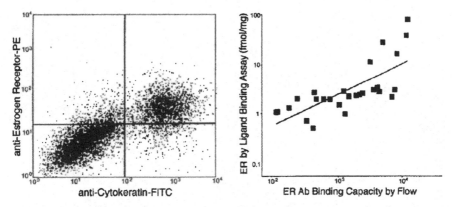

Fig. 7.3. A dot plot (on the left) showing the staining of cells from a human breast tumor for two intracellular proteins. Cytokeratin-positivity marks tumor cells in the suspension, and estrogen receptor positivity on these cells indicates superior prognosis. The plot on the right shows a correlation (in 27 breast tumors) between the intensity of estrogen receptor staining by flow cytometry and the level of estrogen receptor binding (by radioligand binding assay). Modified from Ian Brotherick et al. (1995).

Having discussed the staining of cells for both extracellular and intracellular proteins, and, in the process, learned something about general flow cytometric methodologies for analysis of data, we are now ready, in the next chapter, to apply some of these general methods to cellular components that are not proteins at all.

FURTHER READING

Chapters 12 and 13 in **Darzynkiewicz**, Chapter 15 in **Stewart and Nicholson**, and Chapter 10 in **Bauer et al.** are all good discussions of intracellular staining.

8

Cells from Within:
DNA in Life and Death

In the previous chapters, we have discussed how it is possible to stain proteins on the surface and inside of cells and then to analyze these cells for the presence and intensity of that stain. In addition to protein, another biochemical component that can be used to classify different types of cells is, of course, DNA. It should therefore come as no surprise that flow cytometrists have developed methods for analyzing DNA content.

FLUOROCHROMES FOR DNA ANALYSIS

By comparison with the fluorochromes used for conjugation to antibodies for staining the proteins of cells, DNA-specific fluorochromes have important differences. In particular, whereas fluorescein, phycoerythrin (PE), and others are fluorescent whether or not they are bound to cells, the DNA fluorochromes fluoresce significantly only when they are bound to their target molecules. In addition, unlike the tight binding of antibody to antigen, DNA fluorochromes are generally in loose equilibrium between their bound and free states. Therefore procedures for analyzing the DNA content of cells involve sending cells through the flow cytometer without washing them to remove the "unbound" fluorochrome. The unbound fluorochrome will not add to background fluorescence because it is hardly fluorescent unless bound to nucleic acid. And washing would, in any case, lower overall specific fluorescence by removing much of the fluorochrome (both bound and unbound) from the cell.

TABLE 8.1. Characteristics of Some Nucleic Acid Stains

Stains	Absorption (nm)	Fluorescence (nm)	Specificities
Hoechst 33342 and Hoechst 33258	346	460	DNA with AT preference; Hoechst 33342 enters viable cells well, Hoechst 33258 less well
DAPI	359	461	DNA with AT preference; slightly permeant to viable cells
Chromomycin A3	445	575	DNA with GC preference; impermeant
Acridine orange	460 (RNA) 480 (DNA)	650 (RNA) 520 (DNA)	DNA and RNA; metachromatic; permeant to viable cells
Thiazole orange	509	525	DNA and RNA; permeant
Ethidium bromide	510	595	Double-stranded nucleic acids; impermeant
Propidium iodide	536 (also UV)	623	Double-stranded nucleic acids; impermeant
7-Amino-actinomycin D (7-AAD)	555	655	DNA and RNA; GC preference; impermeant to viable cells
TO-PRO, TO-TO, PO-PO, PO-PRO, YO-YO, and YO-PRO series	434–747	456–770	DNA and RNA; impermeant
SYTO series	488–621	509–634	DNA and RNA; permeant

Several types of fluorescent stain are available for the analysis of DNA; their characteristics make them suitable for different applications (Table 8.1). The most specific stains (e.g., DAPI and the Hoechst dyes, which stain specifically for AT groups on DNA) require the use of a laser with significant ultraviolet (UV) output. Hoechst dyes as well as a newly developed far-red dye called DRAQ5 (alone of all the current DNA-specific stains) also penetrate the outer membrane of living cells and can therefore be used for staining living cells with different DNA content for subsequent sorting for separate culture or functional analysis. Chromomycin A3 is specific for the GC bases in DNA and therefore is an appropriate stain for use in conjunction with Hoechst 33258, as will become evident in the discussion of chromosome techniques. Propidium iodide, although not very specific (it stains all double-stranded regions of both DNA and RNA by intercalating between the stacked bases of the double helix) and not able to penetrate an intact cell membrane, has the decided advantage of absorbing 488 nm light and then fluorescing at wavelengths above 570 nm. This means that, in the presence of RNase, propidium iodide can be used as a DNA stain in cytometers with low-power argon lasers. Propidium iodide has therefore become the most common DNA fluorochrome for flow analysis.

More recently, a series of nucleic acid probes has been developed by Molecular Probes (Eugene, OR); these probes have an array of unlikely names (like TO-PRO, YO-YO, and PO-PRO, sounding something like the three little maids from school in "The Mikado") and also provide a large choice of absorption, emission, and nucleic acid binding properties. Other fluorochromes that absorb 488 nm light include acridine orange, which is metachromatic; that is, it fluoresces red if bound to nonhelical nucleic acid (e.g., RNA or denatured DNA) and fluoresces green if bound to helical nucleic acid (e.g., native DNA). Acridine orange has been used effectively by Darzynkiewicz and coworkers to follow the changes in RNA content and in DNA denaturability that occur during the cell cycle. Moreover, the monoclonal antibody against bromodeoxyuridine (a thymidine analog) can be conjugated to fluorescein, and it will then stain DNA that has incorporated bromodeoxyuridine when cells have been pulse-fed with this compound during DNA synthesis. Before discussing the uses of these stains for chromosome and cell cycle analysis, we should first consider the most obvious use of DNA fluorochromes: staining cells for their total DNA content.

PLOIDY

The amount of DNA in the nucleus of a cell (called the 2C or *diploid* amount of DNA) is specific to the type of organism in question.

Different species have different amounts of DNA in their cells (e.g., human cells contain about 6 pg of DNA per nucleus; chicken cells, about 2.5 pg of DNA per nucleus; corn [*Zea mays*] nuclei, about 15 pg; and *Escherichia coli*, between 0.01 and 0.02 pg each). However, within the animal kingdom, with three major exceptions, all healthy cells in a given organism contain the same amount of DNA. The three major exceptions are, first, cells that have undergone meiosis in preparation for sexual reproduction and therefore contain the 1C or haploid amount of DNA typical of a gamete; second, cells that are carrying out DNA synthesis in preparation for cell division (mitosis) and therefore for a short period contain between the 2C amount of DNA and twice that amount; and third, cells that are undergoing apoptosis and have begun to loose pieces of fragmented DNA. (There are other less common exceptions as well: For example, liver cells exist as normal tetraploids, and multiploidy is the rule rather than the exception in plant cells.) Because healthy, normal animal cells from a given individual, with these three major and other minor exceptions, contain the same amount of DNA, measurement of the DNA content of cells can be used to identify certain forms of abnormality. More specifically, the type of abnormality commonly termed *malignancy* is often associated with genetic changes, and these genetic changes may sometimes be reflected in changes in total DNA content of the malignant cell.

It is possible to permeabilize the outer membrane of normal cells (with detergent or alcohol) in order to allow propidium iodide to enter the nuclei. If we then treat the normal cells with RNase in order to ensure that any fluorescence results from their DNA content (without a contribution from double-stranded RNA), we find that the nuclei fluoresce red with an intensity that is more or less proportional to their DNA content. By the use of a red filter and a linear amplifier on the photomultiplier tube, we can detect that red fluorescence. The channel number of the fluorescence intensity will be proportional to the DNA content of the cells. The method is simple and takes about 10 minutes. Flow cytometric analysis of the red fluorescence from the particles in this preparation of nuclei from normal, nondividing cells will result in a histogram with a single, narrow peak (see the first histogram in Fig. 8.1); all the particles emit very nearly the same amount of red fluorescence. This supports our knowledge that all

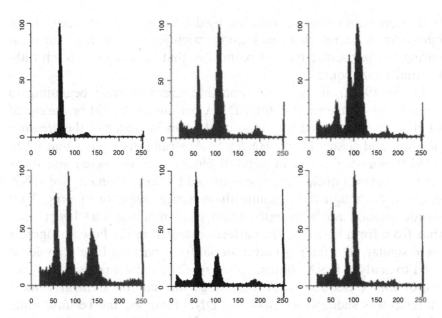

Fig. 8.1. Propidium iodide fluorescence histograms from nuclei of cells aspirated from normal tissue (upper left) and malignant breast tumors. Data courtesy of Colm Hennessy.

normal, nondividing nuclei from any one organism contain the same amount of DNA.

If we then look at a preparation of material from malignant tissue, we find that the fluorescence histogram often indicates the presence of cells with the "wrong" amount of DNA, as well as cells with the amount of DNA that is normal for the organism in question. The normal cells are said to be *euploid* or normal diploid, and the abnormal cells are termed *aneuploid* or *DNA aneuploid* (flow cytometrists have hijacked these terms from cytologists and use them to refer to total DNA content of cells; cytologists feel that the use of the euploid/ aneuploid classification is ambiguous unless chromosomes have been counted). Histograms from examples of some malignant tissues are shown in Figure 8.1. The abnormal peak or peaks may have more or less DNA than normal cells (hyperdiploid or hypodiploid). Because our basic axiom is that all normal cells from an organism contain the same amount of DNA, any tissue that yields a DNA flow histogram with more than one peak contains, by definition, abnormal cells. Flow cytometry is therefore a quick and straightforward method for measuring the particular type of pathology that results in cells with abnormal DNA content.

In the 1980s, at the same time that scientists were beginning to realize that changes in the total DNA per nucleus could be measured by flow cytometry and that this could be an indicator of the presence of abnormal tissue, David Hedley in Australia discovered that when fixed tumors embedded in paraffin blocks were de-waxed and rehydrated, released nuclei could be analyzed by flow cytometry for DNA content. Although the absolute fluorescence intensity of propidium iodide–stained nuclei released from fixed material was lower than that from fresh material, the patterns revealed in the flow histograms were similar. The finding that material from paraffin blocks could be used to analyze DNA content (ploidy) of the individual cells had two important consequences. A very large amount of archival clinical material was suddenly amenable to DNA analysis, and because some of the archival material was 5, 10, and 20 years old, long-term clinical follow up of the patients was immediately available.

The correlation of DNA flow histograms with prognosis became a quick and simple proposition. As a result of Hedley's technique, the corridors in hospitals all over the world were suddenly filled with swarms of young clinicians beating paths to the doors of their pathology departments (and then on to the flow cytometry facilities). An enormous number of publications emerged from the use of this technique on many different types of human material. Aware of the risk of overgeneralization and without the time or space in this book for a full discussion of clinical correlations, I can probably safely say here that most (but not all) of the published results showed a correlation between abnormal flow histograms (aneuploidy) and unfavorable prognosis. Furthermore, many of the publications also showed that flow histograms provide information about prognosis over and above that provided by other, more traditional prognostic indicators.

Although the use of DNA flow histograms to diagnose aneuploidy is both easy and rapid, it does have certain drawbacks that should be made clear. The first drawback results from the nature of malignant changes themselves: Not all malignancies will result from DNA changes that are detectable by a flow cytometer. Current knowledge of the causes of malignancy is far from perfect. Nevertheless, it is possible to imagine that some malignancies may result from changes that are not related to DNA, and other malignant changes may affect a cell's DNA but not in a way that could ever be detected in a flow histogram of propidium iodide fluorescence. For example, chromosome translocations may lead to gross abnormalities in genetic coding, but do not lead to any change at all in the total DNA content of a nucleus. Translocations can be detected easily by microscopic analysis of the individual banded chromosomes in a mitotic spread, but translocations will never be detected by flow cytometry of nuclei stained with propidium iodide. Whereas extra copies (e.g., trisomy) of a large chromosome may result in a measurable shift in the total DNA content of a nucleus, trisomy of a small chromosome may not be detectable in this type of flow analysis (a large chromosome might contain 4% of a cell's total DNA, but a small chromosome has less than 1%). Similarly, small insertions or deletions to chromosomes may lead to changes in DNA content that are too small to be detected by flow cytometry. Any change resulting in less than 3–5% difference in total DNA content may be difficult to detect by flow cytometry, although, to a geneticist, a 3% deletion or insertion involves a large number of base pairs with a potentially enormous amount of misplaced genetic information.

While this first kind of difficulty is an intrinsic limitation of the DNA flow histogram technique, a second problem is more in the nature of a continuing question about interpretation. Although Figure 8.1 shows examples of histograms that provide undoubted evidence of abnormality, Figure 8.2 shows another series of histograms that are considerably more difficult to interpret because of problems arising with so-called wide coefficient of variation (CV) data. The real question concerns our ability to rule out the existence of near-diploid abnormalities when the width (CV) of a peak is very broad. In theory, because all normal nuclei contain the same amount of DNA, the peak in a flow histogram of normal cells should have the width of only a single channel (all the particles should have the same fluorescence intensity and should appear in the same channel). In practice, because staining and illumination conditions will not be exactly uniform, the fluorescence intensities of normal nuclei stained with propidium iodide have a certain range of values. One of the ways in which otherwise quite civilized flow cytometrists compete with each other is by bragging about the small CVs on the peaks of their DNA histo-

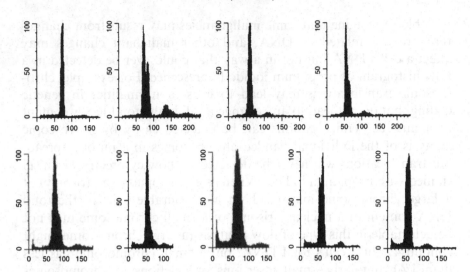

Fig. 8.2. Compared with the narrow peak in the normal histogram at the upper left, it can be seen that a single peak with a wide coefficient of variation (CV) or skewed profile may mask a near-diploid malignant cell line. In addition, an extra small peak at the 4C position may result from clumping of nuclei, cycling cells, or a true tetraploid abnormality. Data courtesy of Colm Hennessy.

grams (as a rule of thumb, <3% is good; >8% is not good). A wide CV might result from old and partially degraded material, from erratic flow due to a partially clogged flow orifice, from a fluctuating laser beam, from a sample that has been run too quickly (remember that widening of the core diameter within the sheath stream may lead to unequal illumination as particles stray from the center of the laser beam), from nuclei that have been unequally exposed to stain, or, finally, from abnormal cells with a DNA content quite close to that of the normal material. The sensitivity of the technique for detecting these near diploid abnormalities and thus for classifying tissue as euploid or aneuploid therefore depends on the cytometrist's ability to obtain narrow CVs in the normal controls.

Another problem concerning interpretation arises from the inconvenient fact that aneuploid tumors often have DNA content that is very close to double the amount found in normal cells. This amount is referred to as *4C* or *tetraploid*. If we stop and think, we can immediately see why this might lead to problems in flow analysis (see the small 4C peaks in Fig. 8.2). First of all, perfectly normal cells with the 4C amount of DNA appear at certain phases in the cell cycle (just before cell division); therefore, if normal dividing cells are present, a significant number of particles may have double the 2C amount of DNA and will therefore appear in a peak at the tetraploid position. Second, remember that the flow cytometer is poor in its ability to distinguish large particles from clumped particles. It is not surprising, then, that the cytometer is, in the same way, inadequate at distinguishing a nucleus with double the normal amount of DNA from two normal nuclei clumped together. Scientists and clinicians usually resort to adopting some threshold value for classification purposes. For example, a sample with a 4C peak may be considered aneuploid only if the tetraploid peak contains more than 10% of the total number of nuclei counted; otherwise it will be considered normal on the assumption that about 10% of normal cells may appear in the tetraploid position owing to clumping and/or mitosis.

CELL CYCLE ANALYSIS

As mentioned above, normal cells will have more DNA than the 2C amount appropriate to their species at times when they are preparing for cell division. The cell cycle has been divided into phases (Fig. 8.3). Cells designated as being in the G0 phase are not cycling at all; cells in G1 are either just recovering from division or preparing for the initiation of another cycle; cells are said to be in S phase when they are actually in the process of synthesizing new DNA; cells in the G2 phase are those that have finished DNA synthesis and therefore possess double the normal amount of DNA; and cells in M phase are in mitosis, undergoing the chromosome condensation and organization that occur immediately before cytokinesis (resulting in the production of two daughter cells, each with the 2C amount of DNA). A DNA flow histogram provides a snapshot of the proportion of different kinds of nuclei present at a particular moment. If we look at the DNA content of cells that are cycling (not resting), we will find some nuclei with the 2C amount of DNA (either G0 or G1 cells), some nuclei with the 4C amount of DNA (G2 or M cells), and some nuclei with different amounts of DNA that span the range between these 2C and 4C populations (Fig. 8.4). A theoretical histogram distribution would look like Figure 8.5. Figure 8.6 shows an example of

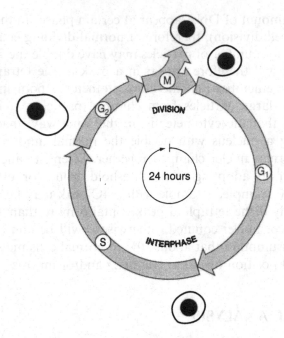

Fig. 8.3. The four successive phases of a typical mammalian cell cycle. From Alberts et al. (1989).

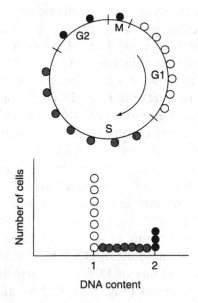

Fig. 8.4. Schematic illustration of the generation of a DNA distribution from a cycling population of cells. From Gray et al. (1990).

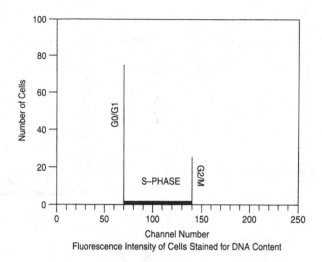

Fig. 8.5. The theoretical histogram generated from the sampling of a population of cycling cells.

DNA flow histograms that result from the propidium iodide staining of cells taken from a culture before and after they have been stimulated to divide.

The traditional method for analyzing cell division involves measuring the amount of DNA being synthesized in a culture by counting the radioactivity incorporated into DNA when the dividing cells are given a 6 h pulse with tritiated thymidine. The DNA histogram resulting from flow cytometric analysis offers an alternative to this technique. By dividing the histogram up with four markers, we can delineate nuclei with the 2C amount of DNA, those with the 4C amount of DNA, and those with amounts of DNA between the two delineated regions and therefore caught in the process of synthesizing DNA. The nuclei making DNA and showing up between the two peak regions should in some way correlate with the values obtained for DNA synthesis based on the uptake of tritiated thymidine. The values are not directly convertible one to the other: The radioactive method reflects the *total amount* of DNA being synthesized and will give higher values when more cells are present, whereas the flow method measures the *proportion* of cells that are in the process of making DNA and will not be affected by increases in the total number of cells. In addition, the radioactive method will give higher values if there is a significant amount of DNA repair going on,

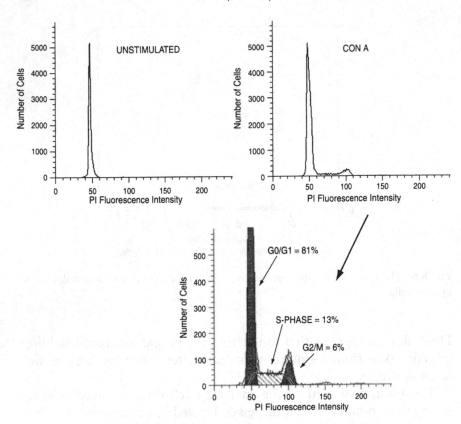

Fig. 8.6. DNA histograms from lymphocytes stimulated to divide.

whereas the flow method will give higher values if a proportion of cells are blocked in S phase. However, with these provisos, flow cytometry does offer a rapid and painless (nonradioactive) method for looking at cell proliferation.

Having agreed on the general principle that flow cytometry in conjunction with propidium iodide staining is an appropriate technology for analyzing cell proliferation, we have now to face the problem that the actual histogram has a certain width to the G0/G1 and to the G2/M peaks and does not look like our theoretical distribution; we have to decide where to place those four markers mentioned above so as to delineate correctly the three regions (2C, 4C, and S phase). In a scenario that may by now be familiar, what seemed like a straightforward question turns out to have a less than straightforward answer. Because the 2C and 4C peaks in a flow histogram have finite widths (remember the discussion about CV in the section on ploidy), it turns out to be rather difficult to decide where the 2C (or G0/G1) peak ends and nuclei in S phase begin. Similarly, it is difficult to know exactly where the distribution from nuclei in S phase ends and the spread from nuclei in G2 or M (4C amount of DNA) begins. In fact, there is no unambiguously correct point to place markers separating these three regions: The regions overlap at their extremes as a result of the inevitable nonuniformity of staining and illumination. The question therefore becomes not where to place the markers delineating the three cell cycle regions, but how many of the nuclei lurking under the normal spread of the 2C and 4C regions of the histogram are actually in S phase. Enter the mathematicians.

Algorithms based on sets of assumptions about the kinetics of cell division and the resulting shape of cell cycle histograms can be used to derive formulae for separating the contribution to the fluorescence distribution from our three separate cell cycle components. The algorithms range from the simple to the complex. They all seem to work reasonably well (that is, they all give similar and intuitively appropriate answers) when cell populations are well behaved. However, they all reflect the intrinsic limitations of using simplistic mathematical models for complex biological systems when cell populations grow too rapidly, are blocked in the cycle, or are otherwise perturbed. Bearing these limitations in mind, we can now look at four of the models used.

Figure 8.7 shows a DNA histogram derived from the propidium iodide staining of cells from a dividing culture. The simplest method for analyzing this histogram is the so-called *peak reflect method* whereby the shape of the G0/G1 peak is assumed to be symmetrically distributed around the mode. Given this assumption, the width of the peak, from the mode to the left (low fluorescence) edge is simply copied to the right (high fluorescence) edge; the same thing is done in reverse with the G2/M peak. Then everything in the middle between these two delineated regions is considered to be the result of S-phase cells.

A slightly more complex method for estimating the proportion of S-phase cells is called the *rectangular approximation method*. This method assumes that cells progress regularly through S phase and therefore that the proportion of cells at any given stage of DNA

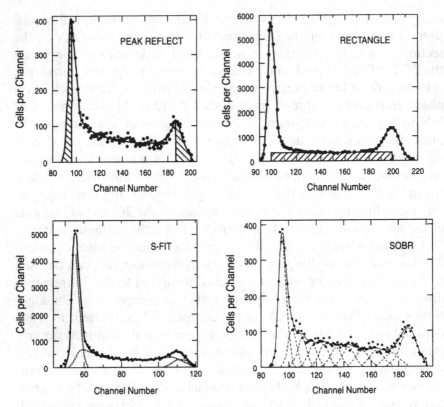

Fig. 8.7. Different mathematical algorithms for determining the contribution of S-phase nuclei to a DNA flow histogram. Upper left and lower right from Dean (1987); upper right and lower left from Dean (1985).

synthesis is constant. When this method is used, the average number of cells in the middle region of the DNA histogram is evaluated, and the height of this region is then extrapolated in both directions, toward the 2C peak and toward the 4C peak. The rectangle derived from this evaluation is then ascribed to S-phase cells, and all the other cells are considered either G0/G1 or G2/M depending on whether they have higher or lower fluorescence than the middle point of the distribution.

The so-called S-FIT method and the sum-of-broadened-rectangles (SOBR) method both use more sophisticated mathematical assumptions to model the shape of the S-phase region of the histogram. A polynomial equation (S-FIT) or a series of broadened Gaussian distributions (SOBR) is derived that best fits the S-phase region of the histogram; then this derived shape is extrapolated toward the 2C and 4C peaks to estimate the contribution of S-phase cells within these regions.

One of the essential problems in assigning cells in a flow histogram to certain stages of the cell cycle is that an aggregate or clump of two G0/G1 cells will have double the normal DNA content (remember our discussion of the problem in diagnosing tetraploid tumors) and will appear as if they are a single G2/M cell. One way of diagnosing a clumped sample is by looking for peaks at the 6C position (resulting from three nuclei together). If there are clumps of three cells, then, statistically, there will be even more clumps of two cells. DNA analysis software can estimate the doublet contribution to the tetraploid peak by using probability algorithms to extrapolate from the triplet peak at the 6C position. The software (Fig. 8.8, right hand plot) can then subtract out this doublet contribution and give a measure of the "true" number of G2/M cells.

As well as software-based aggregate subtraction, so-called pulse processing (or, more recently, digital) electronics can give us help in this task. In general, particles (cells or nuclei) give out signals that

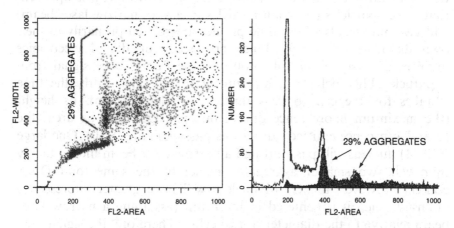

Fig. 8.8. By looking at the peak at the 6C and 8C positions (aggregates of three and four cells), software algorithms use this information to estimate the contribution of clumps of two cells to the peak at the G2/M (4C) position. The graph at the right indicates the software estimation of these aggregated cells. The graph at the left indicates where these cells fall on a plot of signal area versus signal width.

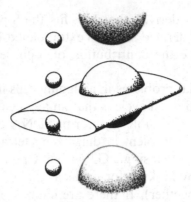

Fig. 8.9. The time that a fluorescence signal lasts (the signal width) depends primarily on the size of the laser beam (if the cell is smaller than the beam) or primarily on the size of the cell (if the cell is larger than the beam). From Peeters et al. (1989).

last, in time, just as long as it takes for the entire particle to move through the laser beam. What this means is that particles with diameters smaller than the laser beam all give out signals that last approximately the same length of time (dependent primarily on the laser beam width in the direction of flow and on the stream velocity); this is a traditional method of flow analysis. However, it is apparent that larger particles (or small particles in a very narrow laser beam) will give out signals whose time profiles are related primarily to their own diameter (Fig. 8.9). Pulse processing or digital electronics involves analysis of the full profile of the fluorescence signal from a particle. This includes a measure of the signal's width (the time it takes for the cell to pass through the laser beam), its height (the maximum fluorescence during this passage), and its area (the total fluorescence emitted during this passage) (Fig. 8.10). One large (G2/M) nucleus will pass through a narrow laser beam more quickly than will two smaller aggregated nuclei of the same total DNA content; however, the G2/M nucleus will have greater fluorescence intensity when it is centered in the beam (assuming a narrow laser beam relative to the diameter of two cells). Therefore, the signal area is used as the DNA parameter in cell cycle analysis because it is most closely proportional to total DNA content of a cell, but the width and height characteristics of the resulting fluorescence signal can be stored as extra parameters and can be used to distinguish clumps

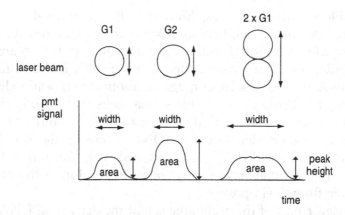

Fig. 8.10. Signal width, area, and height characteristics of light pulses as G1, G2, and two clumped G1 cells move through a narrow laser beam. Modified from Michael Ormerod.

from single G2/M particles. This can help considerably in the interpretation of DNA histograms (Fig. 8.11).

Software algorithms for estimating the proportion of S-phase cells are approximations. They can be refined mathematically in the hope of better approaching the biological truth. A mathematical model will always, however, have trouble coping with a biological situation that is disturbed or contains mixed populations behaving in erratic

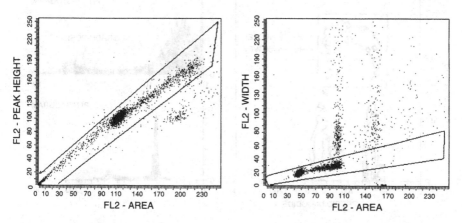

Fig. 8.11. Single cells (shown in the gates) can be distinguished from aggregates because single cells have lower signal widths and greater signal heights relative to their signal areas. The left plot is of data acquired on a Beckman Coulter cytometer; data in the right plot were acquired on a Becton Dickinson cytometer.

ways. Flow cytometry does, however, offer a more direct way to measure DNA synthesis. Bromodeoxyuridine (sometimes abbreviated BrdU or BUdR or BrdUdr) is a thymidine analog. If cells are pulsed with BrdU, it will be incorporated into the cell's DNA in the place of thymidine. Fluorescein-conjugated monoclonal antibodies with specificity for BrdU are available so that cells that have been pulsed with BrdU for a short period of time (about 30 min) can then be treated to partially denature their DNA, exposing the BrdU within the double helix so that it can be stained with the anti-BrdU antibody. Any cells that have incorporated BrdU during the pulse will then stain fluorescein positive.

The clever part of this technique is that the denatured DNA can be stained with propidium iodide at the same time. The resulting two-color contour plots look like those in Figure 8.12. The red fluorescence axis shows the propidium iodide distribution (proportional to DNA content) with which we have grown familiar; the green fluorescence axis shows which of these nuclei have actually incorporated BrdU during the pulse. As might be expected, the cells in the middle region of the propidium iodide distribution have all incorporated BrdU; but a proportion of the cells at either end of the propidium iodide distribution have also done so. This method, while somewhat

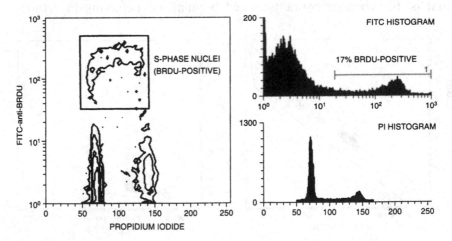

Fig. 8.12. Fluorescein (FITC) histogram, propidium iodide (PI) histogram, and dual-color correlated contour plot of human keratinocytes cultured for 4 days, pulsed with BrdU, and then stained with FITC–anti-BrdU and PI. Data courtesy of Malcolm Reed.

time consuming and a bit tricky technically, does allow a flow cyto-metrist to quantify the proportion of cells in S phase in a way that cannot be done accurately with simple propidium iodide staining. The bromodeoxyuridine method is, in fact, more comparable than simple propidium iodide staining to the traditional method of measuring cell division by assaying the incorporation of tritiated thymidine. It also provides the ability to distinguish cells that may be blocked in S phase from those that are actually incorporating nucleotides.

BrdU staining has also been used to provide information about the kinetics of cell cycles. If we consider cells pulsed for a short period of time with a small amount of BrdU and then killed immediately and stained with both propidium iodide and with a fluorescein–anti-BrdU monoclonal antibody, the fluorescein–antibody should stain equally all cells in S phase (stretching from the 2C peak to the 4C peak). If, however, we wait for some time before killing the cells (and if the BrdU has been used up quickly), then the cells that have incorporated the BrdU (that is, all the cells that were in S phase at the time of the pulse) will have synthesized more DNA, and some of those cells will have progressed into the G2 or M phase of the cell cycle (or indeed cycled back to G1). In addition, some new cells will have started to make DNA after the BrdU had been used up, and these cells will now be in S phase but will have DNA that does not contain BrdU. The contour plots for these cases look like those in Figure 8.13. We can estimate the rate of movement of the BrdU-containing cells through S phase and into the G2 peak by assuming that they are evenly distributed throughout S phase at the time of pulsing and then sampling and staining the cells at one subsequent time. The rate of increase in propidium iodide intensity of the fluorescein-positive nuclei is equivalent to their rate of DNA synthesis and provides us with information about the cycle time of the actively dividing cells. Moreover, the cycle time of the fluorescein-positive cells, in conjunc-tion with the proportion of cells in S phase, can be used to estimate the doubling time of a population of cells.

Given the possible ambiguities in attempting to correlate clinical prognosis with aneuploidy and given the knowledge that malignant cells typically have "out-of-control" or unregulated proliferation, considerable work has been done in an attempt to use flow cytometry to correlate the percentage of cells in S phase with clinical prognosis in malignant disease. Although flow cytometrists tend to have reser-

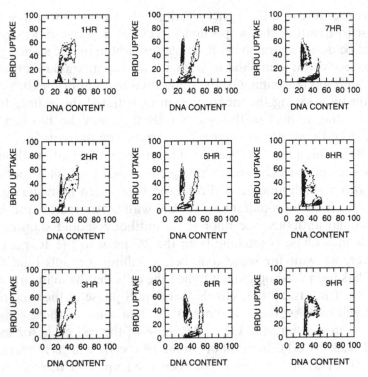

Fig. 8.13. Dual-color distribution of BrdU versus DNA content for Chinese hamster cells pulsed with BrdU and then sampled hourly. From McNally and Wilson (1990).

vations about the reproducibility of S-phase determinations from mathematical modeling of a propidium iodide fluorescence histogram (for the reasons mentioned above), many reports have been published in the medical literature showing useful correlation of the proportion of cells in S phase (the "S-phase fraction") with poor clinical prognosis (see Chapter 10).

TWO-COLOR ANALYSIS FOR DNA AND ANOTHER PARAMETER

Having described the analysis of cells for total DNA content (with propidium iodide) and newly synthesized DNA (with BrdU), we may now go on to consider some other techniques for exploiting the multi-

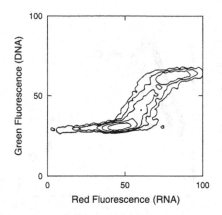

Fig. 8.14. DNA (green fluorescence) versus RNA (red fluorescence) content of human leukemic cells stained with acridine orange. From Darzynkiewicz and Traganos (1990).

parameter potential of the flow cytometer with the dual staining of cells for both DNA and some other parameter. Zbigniew Darzynkiewicz and coworkers in New York have used acridine orange with great success to look at the DNA and RNA contents of cycling cells. Because acridine orange fluoresces red when it binds to RNA and green when it binds to DNA, cells can be examined simultaneously for both constituents. Dual analysis of this type has given us increased information about the progression of cells through the cell cycle. Specifically, it can be seen quite clearly from plots like that in Figure 8.14 that some of the cells with the 2C amount of DNA (G0 or G1) contain increased levels of RNA. The synthesis of RNA appears to be an early event in the entrance of a cell into the division cycle. After this initial increase in RNA content, cells synthesize more RNA as they begin to synthesize DNA (S phase). At mitosis, they have increased levels of both RNA and DNA compared with resting cells. Analysis of the acridine orange staining of cycling cells has led, in particular, to improved ability to analyze early events in the division cycle.

This acridine orange technique has been taken one step further by Darzynkiewicz's group. Because acridine orange fluoresces red when bound to single-stranded nucleic acid, but green with double-stranded nucleic acid, acridine orange can be used under mildly denaturing

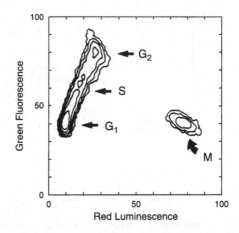

Fig. 8.15. L1210 cells treated with RNase and acid and then stained with acridine orange reveal that DNA in mitotic cells is extensively unwound and exhibits increased red and decreased green fluorescence relative to DNA from cells in other phases of the cell cycle. From Darzynkiewicz (1990).

conditions (and in the presence of RNase) to distinguish easily denatured from more resistant forms of DNA. Once RNase has been used to remove the RNA, mild denaturation can be used to cause unwinding of the helical forms of DNA that are loosely packed (e.g., in cells that are in the process of DNA synthesis); it will not affect DNA that is tightly condensed (e.g., in resting cells). Figure 8.15 indicates the kind of contour plots that can be obtained from this type of procedure.

Dual staining of cells can also be used to look at DNA and protein markers simultaneously. Using methods similar to those described in the previous chapter for looking at cytoplasmic proteins simultaneously with surface membrane proteins, cells can be stained for surface proteins, then fixed and permeabilized, and then stained for DNA. Figure 8.16 shows the way that this type of technique can be used to permit the cell cycle analysis of subpopulations of cells independently. In this particular example, it can be seen that, after treatment with the mitogen phytohemagglutinin, it is the CD8-positive cells more than the CD8-negative cells that have been induced to enter S phase.

Similar techniques can be used to stain cells for both DNA and internal (cytoplasmic or nuclear) markers. By fixing the cells, then

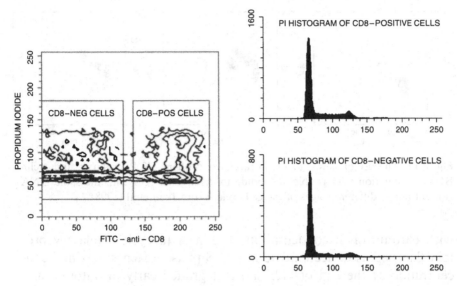

Fig. 8.16. Cell cycle analysis of CD8-positive and CD8-negative cells. Lymphocytes were cultured with phytohemagglutinin, stained with FITC–anti-CD8 monoclonal antibody, treated with saponin to permeabilize the outer membrane, and then stained with propidium iodide (PI) and RNase. Cells provided by Ian Brotherick.

staining for cytoplasmic markers like cytokeratin (as in the example of staining tumor cells for cytokeratin and estrogen receptors in the previous chapter, Fig. 7.3), and then staining with propidium iodide, classes of cells can be selected for ploidy analysis. In this way, for example, breast tumor cells (which are likely to be of epithelial origin) can be gated in a mixed population from a tumor. Then the cells gated for fluorescein–anti-cytokeratin positivity can be further analyzed to see if any of these nuclei are aneuploid. By this technique, minor populations of tumor cells within a heterogeneous population may be selected and their ploidy determined with more sensitivity.

The Darzynkiewicz laboratory has continued, more recently, to use flow cytometry in creative ways to contribute to our knowledge of the regulation of cell division. By staining for DNA as well as cell cycle-regulatory proteins (the cyclins), the presence of these proteins can be associated with certain stages of the cell cycle. Figure 8.17 shows an example of data derived from a protocol staining cells for DNA content along with antibodies against the nuclear proteins H3-P (H3 is a histone, which, when phosphorylated, is associated

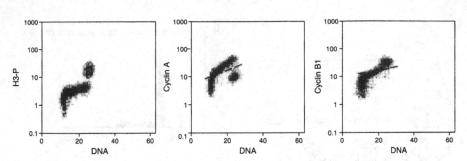

Fig. 8.17. The use of stains for the intracellular proteins H3-P, cyclin A, and cyclin B1 in conjunction with propidium iodide to distinguish cells with the same DNA content but at different stages of the cell cycle. From Juan et al (1998).

with chromosomal condensation), cyclin A (a cycle-regulatory protein that begins to accumulate at mid-S phase, reaches maximal concentration at the end of G2, and is degraded early in mitosis), and cyclin B1 (which begins to accumulate at the beginning of G2 and is maximal at the beginning of mitosis, declining at the cell's entry into anaphase). H3-P can dichotomize 4C cells into those that are in the G2 stage of the cycle and those that are actually mitotic. Similarly, the cyclins can additionally separate cells at different stages of mitosis. Figure 8.18 indicates a summary of the time course for expression of cyclins D, E, A, and B.

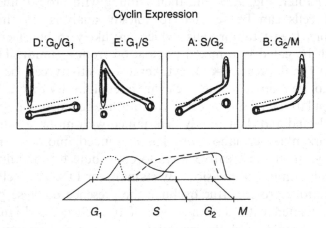

Fig. 8.18. A summary of the time course of expression of cyclins D, E, A, and B through the cell cycle. DNA is plotted on the x-axis of the coutour plots and cyclin expression on the y-axis. Courtesy of James Jacobberger.

Some particular issues arise when combining DNA staining with protein staining; these issues are related to those discussed in the previous chapter on combining the staining of intracellular with extracellular proteins. In both cases, the fixation procedure, while maintaining the integrity of the cells after permeabilization, can also affect the stainability of the molecules in question. In the case of DNA staining, fixation of cells will cross-link histones on the chromosomes, thus limiting the access of DNA-specific fluorochromes to their binding sites. In practice, this means that cells fixed with formaldehyde will maintain their cytoplasmic integrity well but will show decreased propidium iodide fluorescence and wider CVs than will cells fixed with, for example, ethanol. The solution is, once again, a compromise. Short fixation with low concentrations of formaldehyde followed by detergent treatment works well to permit maintenance of cytoplasmic proteins along with fairly good DNA profiles. Exact procedures need to be individualized to the proteins and cells in question.

CHROMOSOMES

Until now, the particles flowing through our flow cytometer have been cells or nuclei. Other types of particles are, of course, possible. One of the best examples of the successful application of flow cytometry to noncellular systems has been in the analysis of chromosomes. In this case, the particles flowing through the system are individual chromosomes that are released from cells that have been arrested in metaphase (much the same conditions as those that are used to prepare chromosomes for analysis in metaphase spreads under the microscope). The released chromosomes are stained with a DNA stain (like propidium iodide) and then sent through the flow cytometer. The resulting histograms of fluorescence intensity reveal peaks whose positions along the x-axis are proportional to the amount of DNA in the chromosome and whose areas are proportional to the number of chromosomes with that particular DNA content (Fig. 8.19). Histograms of this type are called *flow karyotypes*, by analogy with the microscope karyotypes derived from conventional genetic analysis.

Figure 8.20 shows examples of flow karyotypes from different

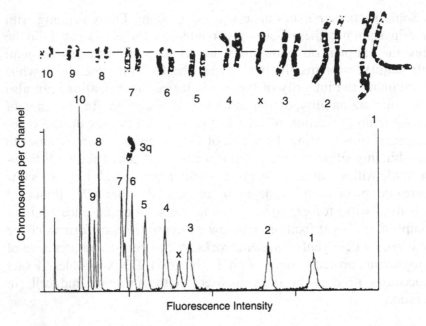

Fig. 8.19. A flow karyotype (fluorescence histogram) of Chinese hamster chromosomes stained with propidium iodide (PI). The G-banded chromosomes from this particular aneuploid cell line are included for comparison with the histogram peaks. From Cram et al. (1988).

species. While some species with small numbers of chromosomes reveal relatively simple histogram patterns, the 23 pairs of chromosomes in the human lead to a rather complex pattern. It is apparent that, although 23 pairs of chromosomes are readily distinguished under the microscope by a combination of size, centromere position, and banding patterns, many of these pairs have similar total DNA content and are not distinguishable in a flow histogram. We can obtain considerable help by using Hoechst 33258 and chromomycin A3 in a dual staining system: Hoechst 33258 stains adenine- and thymine-rich regions of DNA preferentially, and chromomycin A3 is specific for regions rich in the guanine and cytosine base pairs. By using this dual system, we find that some chromosomes with closely similar total DNA content have differing base pair ratios. Compare particularly the positions of chromosomes 13–16 in the one-dimensional histograms (Fig. 8.20A) with these same chromosomes (now separable) in the contour plot (Fig. 8.21). Unfortunately, chromosomes 9–12 are not distinguishable with either system.

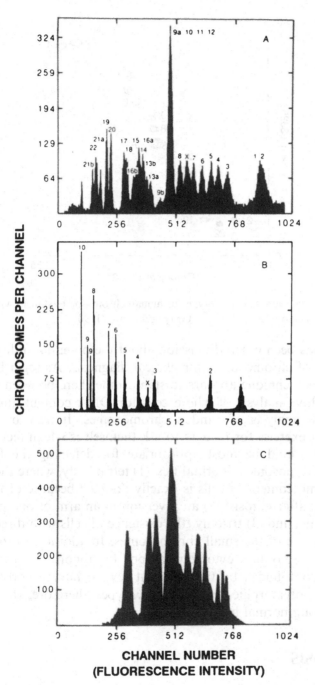

Fig. 8.20. Flow karyotypes from human chromosomes (**A**), hamster chromosomes (**B**), and mouse chromosomes (**C**). From Gray and Cram (1990).

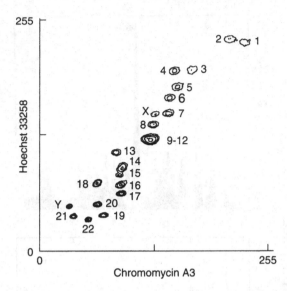

Fig. 8.21. A bivariate flow karyotype for human chromosomes stained with Hoechst 33258 and chromomycin A3. From Gray and Cram (1990).

There has been much discussion about the potential utility of flow cytometry of chromosomes for clinical diagnosis. As regards its sensitivity, this technique appears to stand somewhere between the technique of flow analysis of whole cells for DNA content and that of microscope analysis of banded chromosomes. It may be a useful intellectual exercise for readers to ask themselves which technique or techniques would be most appropriate for detecting the following types of chromosome abnormalities: (1) tetraploidy, where the normal chromosome content of cells is exactly doubled because of failure of cytokinesis after mitosis; (2) an inversion in an arm of one particular chromosome; and (3) trisomy (the existence of cells with three instead of two) of one of the small chromosomes. In addition to these limitations, the use of flow cytometry to look for abnormal chromosomes has been confounded by the fact that several human chromosomes are highly polymorphic, and flow karyotypes, therefore, vary considerably among normal individuals.

APOPTOSIS

We can close this chapter with discussion of two "end of life" applications—apoptosis and necrosis. Although a naïve observer might see death as an unfortunate (and scientifically boring) end-point, recent work indicates that cell death is neither boring nor necessarily unfortunate. Cells often die by an active process that is an important part of the maintenance of organismal homeostasis. This process is called *apoptosis*. Apoptosis, or programmed death, can, for example, prevent the survival of potentially malignant cells with damaged DNA.

Cells that are undergoing apoptosis progress through a series of events. Many of the events that we measure in association with apoptosis, however, may be overlapping in time and therefore may not be a "pathway" so much as symptoms of the underlying process. Some of the events are not obligatory and may differ depending on the nature of the apoptotic trigger and the cell type. In any case, several of these apoptotic-associated events can be analyzed by flow cytometry (Fig. 8.22). One of the events is the flipping and stabilization of phosphatidylserine from the inner surface of the cytoplasmic membrane to the outer surface. On the outer surface, phosphatidylserine appears to identify cells as targets for phagocytosis. Because annexin V binds to phosphatidylserine, staining of intact cells with fluorochrome-conjugated annexin V will detect cells that are in early stages of apoptosis.

A two-color dot plot (Fig. 8.23) of cells stained with propidium iodide and annexin V–FITC will indicate cells in three of the four quadrants. Unstained cells are alive and well and are the double negatives; they neither express phosphatidylserine on their surface nor take up propidium iodide through leaky membranes. Cells that stain just with annexin V are apoptotic; they have begun to express phosphatidylserine on their surface, but have not yet gone through the process that leads to permeabilization of their cytoplasmic membrane. Cells that stain both with propidium iodide and annexin V are necrotic (that is, dead); they take up propidium iodide and also stain with annexin V. With a permeable cell, the flow cytometer cannot tell us whether the annexin V is on the outside of the membrane (because the cells have gone through apoptosis before membrane permeabilization) or on the inside of the membrane (because the cells have died by the necrotic pathway without apoptosis).

Another event associated with apoptosis is the fragmentation of DNA due to endonuclease activation. This fragmentation results in the appearance of increased numbers of "free ends" on the DNA molecules in the cell. By incubating fixed and permeabilized cells with

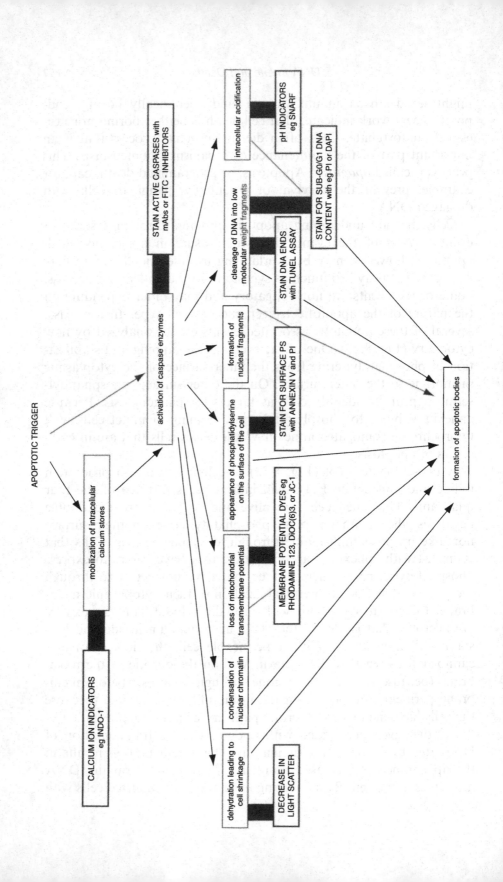

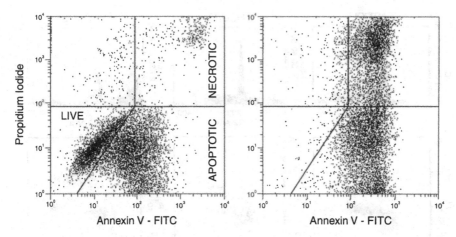

Fig. 8.23. Cells stained with propidium iodide for permeable membranes and with annexin V for phosphatidylserine can distinguish live (double negative), apoptotic (annexin V-positive, propidium iodide–negative), and necrotic (double positive) cells. This figure (with data from Robert Wagner) shows cultured bovine aortic endothelial cells; the left plot displays data from adherent cells, and the right plot displays data from the "floaters."

the enzyme terminal deoxynucleotidyl transferase (TdT) and with fluorochrome-labeled nucleotides, cells with greater numbers of DNA fragments incorporate more of the fluorescent nucleotides onto their increased number of termini. The resulting increased cellular fluorescence is indicative of an apoptotic cell (Fig. 8.24, lower dot plots). Appearance of positive cells in this TdT flow assay correlates well with the classic gel electrophoresis assay for apoptosis, where "ladders" of small DNA fragments are indicators of endonuclease activation. In a desperate attempt to find an acronym, this flow assay for apoptosis has been called the TUNEL assay (for *T*erminal deoxynucleotidyl transferase–mediated d*U*TP *N*ick *E*nd-*L*abeling).

Propidium iodide staining alone can be used to detect later stages of apoptosis (Fig. 8.24, upper histograms). When the DNA becomes extensively fragmented, the small fragments begin to be ejected from

◄ ────────────────────────────────

Fig. 8.22. A diagram of some of the events associated with apoptosis, showing related flow cytometric assays (outlined in bold boxes). Although it is generally agreed that calcium flux and caspase activation are early events following the trigger signal, the events following on from there may be essential steps in an ordered pathway or may be independent and merely symptomatic of apoptosis.

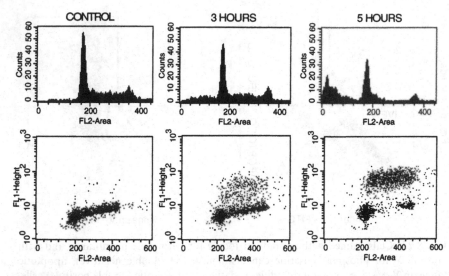

Fig. 8.24. Cells in late apoptosis have fragmented DNA. They can be visualized as cells with increased incorporation of fluorochrome-labeled nucleotides to the ends of the fragments in the flow cytometric TUNEL assay (lower dot plots, with FL1 as FITC-dUTP and FL2 as propidium iodide bound to DNA). Alternatively, they can be visualized as cells with less-than-normal (sub-G0/G1) DNA content (upper histograms of propidium iodide fluorescence). Data from etoposide-treated ML-1 cells courtesy of Mary Kay Brown and Alan Eastman.

the cells and/or can be washed from the cells during the staining protocol. In either case, the apoptotic cell at this stage will have significantly less DNA than a healthy, viable cell. This lowered DNA content can be detected as a "sub-G0/G1" peak in a simple, one-color DNA histogram. It needs to be mentioned that staining for the cells with sub-G0/G1 DNA content requires a protocol that encourages apoptotic cells to release their fragmented DNA. In contrast, the TUNEL assay uses fixation to ensure that the fragmented DNA is retained.

NECROSIS

Unlike apoptosis, which involves the scheduled and active coordination of metabolic processes, necrosis is a passive response to a toxic or injurious environment. Whereas in apoptosis the cell membrane remains intact until very late in the game, permeabilization of the cell membrane is an early event in necrosis. Because propidium iodide

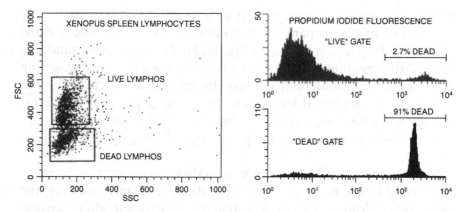

Fig. 8.25. The use of propidium iodide to monitor cell death. Dead lymphocytes have a less intense forward scatter signal than do live lymphocytes. Cultured *Xenopus* spleen lymphocytes courtesy of Jocelyn Ho and John Horton.

is excluded from entering cells by an intact plasma membrane and because it only fluoresces when intercalated between the bases of double-stranded nucleic acid, it will not fluoresce if it is added to a suspension of intact cells. The intact plasma membrane forms a barrier, keeping propidium iodide and nucleic acids apart. It is only when the outer membrane has been breached that the cells will emit red fluorescence. Propidium iodide (or other membrane-impermeant DNA fluorochromes) is therefore a stain (like trypan blue) that can be used to mark necrotic cells (on the reasonable but not necessarily valid assumption that cells with holes in their membranes large enough to allow the penetration of propidium iodide are actually dead according to other viability criteria and vice versa). By this method, cell viability can be monitored in the presence of various cytotoxic conditions (Fig. 8.25).

The only difficulty in using flow cytometry to monitor cell death is that, as mentioned in Chapter 3, dead cells have different scatter properties than living cells. In particular, because of their perforated outer membrane, they have a lower refractive index than living cells and therefore have forward scatter signals of lower intensity. For this reason, it is important not to use a gate or forward scatter threshold when analyzing a population for the proportion of dead and live cells. Any forward versus side scatter gate drawn around normal lymphocytes, for example, will always show most if not all of the cells within that gate to have excluded propidium iodide no matter how many cells in the preparation are dead; this is simply because the dead cells drop out of the gate (Fig. 8.25). The lesson here (and one that needs repeating in reference to most flow analysis) is that gating is an analytic procedure that needs to be performed carefully and with explicit purpose. If we do not define our population of interest (by gating) with care, our results may be accurate answers to inappropriate questions (e.g., "what percentage of living lymphocytes are alive?").

The use of membrane-impermeant DNA fluorochromes to mark dead cells has a more routine application in simply allowing the exclusion of dead cells from flow analysis. This is important because dead cells have the habit of staining nonspecifically when their broken membranes tend to trap monoclonal antibodies directed against surface antigens. Therefore a cell suspension including many dead cells may show high levels of nonspecific stain. By adding propidium iodide to cells just before analysis, the cells fluorescing red can be excluded from further analysis by gating on red negativity and only the living cells then examined for surface marker staining. Figure 8.26 shows the way that this technique can be used when looking at fluorescein staining for surface antigens on cultured mouse thymocytes.

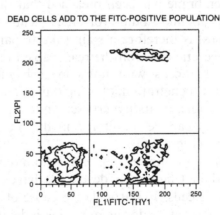

Fig. 8.26. The use of propidium iodide to exclude dead cells from analysis of a mouse spleen cell population for the expression of the Thy-1 surface antigen. The cells were stained with fluorescein for Thy-1 and then with propidium iodide to mark the dead cells. It can be seen that the dead cells (upper right quadrant) contribute to the fluorescein-positive population. Stained cells courtesy of Maxwell Holscher.

The use of propidium iodide or equivalent dyes to exclude dead cells from analysis is a procedure that is recommended as routine for any system staining for surface markers. It is particularly important in analysis of mixed populations where a high percentage of the cells may be dead.

It should, however, be remembered that fixed cells cannot be stained with propidium iodide for live/dead discrimination. Fixed cells are, in fact, all dead and will therefore all take up propidium iodide even if some were alive and some dead before fixation. Ethidium monoazide offers an alternative to propidium iodide if cells will be fixed before flow analysis. It is a dye that, like propidium iodide, only enters dead cells. It has, however, the added advantage of forming permanent cross-links with DNA when photoactivated. Therefore, cells can be stained for surface proteins, incubated with ethidium monoazide under a desk lamp, washed, and then fixed. At the time of acquisition of data on the flow cytometer, red fluorescence will mark the cells that were dead before fixation.

FURTHER READING

Chapters 13–16 and 24 in **Melamed et al.** and Chapters 11–13 in **Watson** are detailed reviews of the applications of flow cytometry to various aspects of cell cycle research.

Several chapters in **Ormerod** and several sections in the **Purdue Handbook**, in **Diamond and DeMaggio**, and in **Current Protocols in Cytometry** give good practical protocols. **Darzynkiewicz** (both the 1994 and 2001 editions) include many chapters with protocols and advice on both the simpler and the more complex methods in nucleic acid, apoptosis, and cell cycle analysis.

Good discussions of the mathematical algorithms for cell cycle analysis can be found in Chapter 6 of **Van Dilla et al.**, Chapter 23 of **Melamed et al.**, and in the multiauthor book Techniques in Cell Cycle Analysis, edited by Gray JW and Darzynkiewicz Z (1987), Humana Press, Clifton, NJ.

The "Cytometry of Cyclin Proteins" is reviewed by Darzynkiewicz Z, et al. (1996) in Cytometry 25:1–13.

Molecular biology and chromosome analysis are discussed in Chapters 25–28 of **Melamed et al.** A special issue (Volume 11, Number 1, 1990) of the journal Cytometry, although somewhat aged, is devoted to the subject of analytical cytogenetics; it contains articles about work at the interface between theory and clinical practice (as well as some beautiful pictures).

A review article by Darzynkiewicz Z, et al. (1997) on "Cytometry in Cell Necrobiology: Analysis of Apoptosis and Accidental Cell Death (Necrosis)" appeared in Cytometry 27:1–20. Boehringer Mannheim produces a free, but substantial manual (now in its second edition) on methods related to the study of cell death.

9

The Sorting of Cells

Although early flow cytometers were developed as instruments that could separate or sort particles based on analysis of the signals coming from those particles, it is now true that most flow cytometry involves analyzing particles but not actually sorting them. This is just one of many examples of the way in which flow technology has moved in unpredictable directions. Many cytometers now do not even possess a sorting capability, and the instruments that can sort particles may be used only infrequently for that purpose. Nevertheless, sorting cells or chromosomes with a flow cytometer is an elegant technology; it may be the only method available for obtaining pure preparations of particles with certain kinds of characteristics of interest. On both of those counts it is certainly worth knowing about sorting even if you are not interested in using the technique at the moment. Now that you are comfortable with flow cytometry and its basic applications, this chapter will present a bit more technical information about the way a sorting flow cytometer (Fig. 9.1) can be used to sort cells.

SORTING THEORY

If a stream of liquid is vibrated along its axis, the stream will break up into drops (pick a nice summer day and try it with a garden hose). The characteristics of this drop formation are governed by an equation that will be familiar to anyone who has studied light waves: $v = f\lambda$, where v is the velocity of the stream, f is the frequency of the vibration applied, and λ is the "wave length" or distance between the drops. One other fact that needs to be known before we can under-

Fig. 9.1. A sorting flow cytometer (MoFlo by Cytomation). The outward complexity of this instrument compared with benchtop cytometers (see Fig. 1.5) reflects the electronic controls necessary for sorting as well as the adaptability of research cytometers with regard to multiple lasers and to the filters and mirrors in the optical light path for multiparameter analysis.

stand sorting is that drops form in a regular pattern so that the distance between drops is, under all conditions, equal to just about 4.5 times the diameter of the stream. For the usual flow cytometer, if a nozzle with a 70 μm orifice is used, the drops will form with a wavelength of about 315 μm. In most cytometers, the stream velocity is set at approximately 10 m/s. Because λ (315 μm) and v (10 m/s) are both now determined, the frequency to achieve drop formation is fixed as well. The nozzle needs to be vibrated at

$$f = (10)/(315 \times 10^{-6}) = 31,746 \text{ cycles per second}$$

to get drops to form (and this kind of restrictive relationship between v, f, and λ is why drops will form from the garden hose only when you get it vibrating at a certain frequency). Figure 9.2 illustrates this relationship between drop generation frequency, stream velocity, and

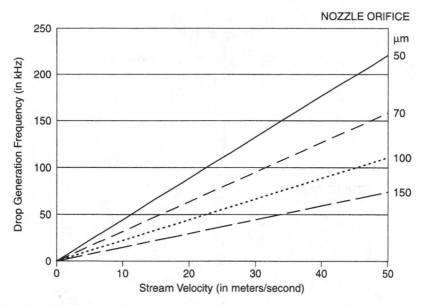

Fig. 9.2. The vibration frequency required to generate drops is determined by the stream velocity and the nozzle orifice diameter. Faster drop generation permits faster cell flow rate without enclosing significant numbers of multiple cells in single drops.

nozzle orifice size. With a flow nozzle vibrating at about 30,000 cycles per second, if the sample is of a concentration such that particles are moving down that stream at 30,000 particles per second, there will be, on average, one particle in every drop; if the particles are less concentrated and flowing at 3000 particles per second, there will be, on average, one particle in every tenth drop.

When a stream vibrates, drops break off from the jet at some measurable distance from the fixed point of vibration. Therefore a particle can be illuminated and its signals detected without interference as it flows within the core of a stream through a vibrating nozzle—as long as the illumination and detection occur close to the point of vibration (Fig. 9.3). At some distance after the point of analysis, the stream will begin to break up into drops; particles will be contained in some of those drops as they break off. It remains now for the cytometer to put a charge on the drops containing particles of interest. Because the drops will flow past positively and negatively charged high-voltage plates, any drop carrying a positive or negative

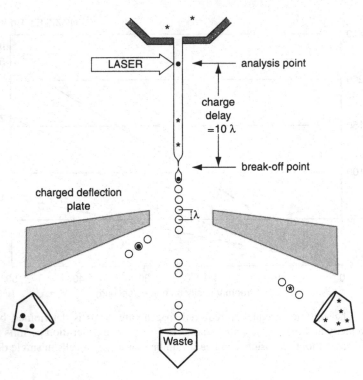

Fig. 9.3. Droplet formation for sorting. A time delay is required between analysis of a particle and the charging of the stream so that only the drop (or three drops) surrounding the desired particle will be charged and then deflected.

charge will be deflected out of the main stream and toward one or the other plate. Once this happens, it is easy to place test tubes, Eppendorf tubes, or wells of a microtiter plate in position to catch the drops deflected to the left or right.

The main trick of flow sorting is, therefore, to apply a charge only to the correct drops (i.e., the ones containing the desired particles) and to no others. To do this, we need to know the time that elapses between the moment a particle is analyzed at the analysis point and the moment, a bit later, that that particle is just about to be trapped in a newly formed drop separating from the stream. If the entire stream is charged (either negatively or positively) by applying a charge at the nozzle just before the drop is formed with the desired particle within it, and then the entire stream is grounded at the nozzle to remove that charge just after the drop in question is formed and has detached from the stream, then only the drop with the desired particle within it will remain charged. Therefore, all we need to do is determine the time it takes a particle to move from the analysis point to the point where the drop containing it breaks away from the stream.

Different methods are available for determining the time delay between illumination of (and detecting signals from) the particle and the trapping of that particle in an isolated drop. In some systems, the operator does a "test sort matrix" of sorted cells or fluorescent beads into puddles on a slide. By changing the timing of the drop charge, the operator can check (with a microscope) to see which time delay produces sorted cells or beads in the puddle. In other systems, we can measure the distance between the analysis point (the laser spot on the stream) and the drop break off point. Then, to convert that distance into a time, we count the number of drops that occur in an equal distance measured further down the stream. Because we can know the frequency of vibration that is driving the nozzle (in cycles per second), we therefore also know the number of seconds per cycle. By counting the number of drops that occur in a distance equal to the distance between analysis point and drop break-off point, we can convert the counted number of drops into a time in seconds. It is that time that is the time delay we want between illumination of the particle and charging the stream. Fortunately, most cytometers will do most of these calculations for us. We simply need to count the number of drops that occur in a distance equivalent to the distance between analysis point and drop break-off point. The cytometer electronics then converts that drop number into a time delay, being the amount of time we need to delay our charging of the stream so that a particle analyzed at the laser intercept and found to have desired characteristics has moved in the stream and is about to be trapped in a forming drop.

Once the charge delay time has been determined (either by counting drops or by a test matrix), the cytometer can then be given sorting regions delineating the flow cytometric characteristics of the desired particles. These regions are, like any flow cytometric regions, simply the numbers (on a scale of 0–1023) that describe the intensity of the light signals characterizing the particles of interest. They can involve single flow parameters (e.g., a particular range of green fluorescence intensity), or they can involve multiple parameters and combinations of regions (e.g., a forward and side scatter region combined with a range of green, orange, and red intensities defining the phenotype of a subset of cells). If we are at all unsure about our accuracy in timing the delay for the drop charge, we can charge the stream for a slightly longer time (centered around the calculated time delay) so that two or three drops are charged after a desired particle has been detected at the analysis point. This should ensure that our desired particle is deflected, even if it should move slightly faster or slightly more slowly than anticipated.

As discussed earlier, with a nozzle of 70 μm and a stream moving at 10 m/s, our system is committed to a vibration frequency of about 30,000 cycles per second (30 kHz) in order to get drops to form. If we prefer a margin of safety, we may want to charge and sort three drops at a time; in that case, we will want a particle in no more than every third drop. This means that our total particle flow rate can be no faster than 10,000 particles per second. Because cells are not spaced absolutely evenly in the flow stream (they obey a Poisson distribution), most sorting operators like to have particles separated by about 10–15 empty drops. With a 70 μm nozzle and a stream velocity of 10 m/s, this restricts our total particle flow rate to about 2000–3000 particles per second. For sorting cells of very low frequency within a mixed population, this may involve unacceptably long sorting times.

The startling implications of this restriction on particle flow rate according to the frequency of drop generation can be seen in Table 9.1. If, in order to separate cells by 15 empty drops, the rate of particles flowing is restricted to 2000 per second, then the particles we

TABLE 9.1. Cell Sorting: Time Needed to Sort Required Number of Desired Particles at Total Particle Flow Rate of 2000 Particles per Second

Required No. of desired particles	Desired particles as % of total particles			
	0.1%	1.0%	5.0%	50.0%
10^3	8.3 min	50 s	10 s	1 s
10^4	1.4 h	8.3 min	1.7 min	10 s
10^5	14 h	1.4 h	17 min	1.7 min
10^6	5.8 d	14 h	2.8 h	17 min
10^7	1.9 mo	5.8 d	1.2 d	2.8 h
10^8	1.6 yr	1.9 mo	12 d	1.2 d

actually desire to purify by the sorting procedure will be flowing at something less than that rate. If the desired particles are 50% of the total, then they will be flowing at a rate of 1000 per second (3.6 million per hour). If, however, the desired particles are only 0.1% of the total, they will be flowing at a rate of 7200 per hour. The time taken for sorting cells, then, depends first on how fast your cells can be pushed through the cytometer without unacceptably high rates of coincidence with multiple cells in drops; second, on what percent the desired particles are of the total number of particles present; and third, on how many desired particles are actually required at the end of the day (or month!). Scientists tend to think logarithmically: It is easy to say that perhaps 10^6 or 10^7 sorted particles are required for a given application. But sorting 10^7 particles takes 10 times longer than sorting 10^6 particles; this can represent a very large investment in time. One moral of this story is that any preliminary procedure that can be applied to enrich a cell suspension before flow sorting will save considerable amounts of time.

Of course, with a little thought it should also be clear that cells could be run through the sorter more quickly without the risk of multiple cells in drops if the drops could be generated at a faster rate. According to our equation, this can be done either by making the nozzle orifice diameter smaller or by running the stream at a higher velocity. The first solution tends to be impractical, as a clogged nozzle ends up being very slow in the long run. The second solution was implemented initially at the Livermore and Los Alamos laboratories, where a high-pressure system (about 100–200 psi) brings about a stream velocity of about 40–50 m/s and a droplet frequency of 150,000–200,000 drops per second. As a result, particles flowing at rates of 10,000–20,000 particles per second can be sorted without significant risk of multiple cells in a drop. Total sorting times can be cut by a factor of 10. High-speed sorting has recently become available on commercial instruments. Modifications of tubing strength, electronics dead time, and air controls have permitted the pressure on the sheath tank to be increased from about 10–15 psi to 30–100 psi. This pressure brings about an increase of the stream velocity from 10 m/s to about 20–40 m/s. Drop drive frequencies need then to be increased (according to the equation) to 63 to 127 kHz (with a 70 μm nozzle) to produce drops. Cells can therefore flow at 6000 to 13,000 cells per second and still appear (on average) in only every tenth drop. If higher abort rates can be tolerated, then these flow rates can be tripled (with one cell in, on average, every third drop).

CHARACTERIZATION OF SORTED CELLS

The first matter of concern after a sort is always purity, but purity is never a straightforward issue (just like many other aspects of flow cytometry). Sorting of test beads (as a bench mark) routinely gives purity of better than 99%. Figure 9.4 shows four real-world examples of cells before and after sorting. In Figure 9.4, the sort regions on the plots in the right column (the sorted cells) have not been moved from where they were set as sort gates before the sort (left column). These plots illustrate that, at times, the accuracy of the sorter may appear to be suspect. Our good results with beads should, however, lead us to conclude that the problem with cells may have nothing to do with the accuracy of the flow cytometer.

The sort may appear to be inaccurate (and the sorted cells less than perfectly pure) because the actual intensity of the cells may decrease between the time a cell is sorted and the time it is run back through

▶

Fig. 9.4. Four examples of cells sorted by Gary Ward in the Englert Cell Analysis Laboratory. The plots in the column on the left are from cells before the sort, indicating the region used to set the sort gate; the column on the right shows the sorted cells, run back through the cytometer for a "purity check." The top row shows mouse B lymphocytes that have been transfected with a gene for the human IgA Fc receptor (a mixed clone cell line from M van Egmond and J van de Winkel; sort experiment by Mark Lang); the cells have been stained for the IgA Fc receptor and then sorted for the top 50% expressors. The second row down shows data from a cell line derived from the tracheal epithelium of a cystic fibrosis patient and invaded by *Staphylococcus aureus* bacteria carrying a plasmid promoter library driving the expression of green fluorescent protein (GFP); GFP expression within the eukaryotic cell (but not in vitro) allows the sorting of cells containing bacteria with promoters activated by intracellular signals (data courtesy of Niles Donegan). The next row down shows monocytes sorted on the basis of their forward and side scatter signals from a preparation of white cells from human peripheral blood (data from Gary Ward). The bottom two rows illustrate data (from James Gorham) for mouse spleen cells (pre-purified with anti-CD4 magnetic beads) stained with anti-CD4–FITC and anti-CD62L–PE (a marker for naïve cells); the two-way sort separated naïve and memory CD4–T cells in order to compare their biological activities in vitro.

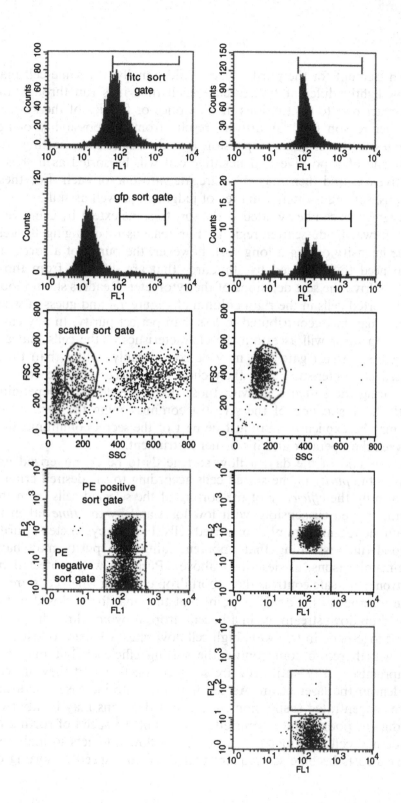

the instrument for the purity check. Additionally, the same cell may show slightly different fluorescence each time it is run through the cytometer due to fluctuations in the optics or fluidics of the system. Another reason for "impurities" results from the possibility of aggregates in the original suspension that dissociate after sorting. An aggregate of a positive and negative cell will be sorted as if it is a positive cell and then may disaggregate into one of each. For these reasons, sorting is often a matter of judgment as well as skill.

Aggregates can be avoided in the sort to some extent by using very tight forward/side scatter regions. For reasons of changing fluorescence intensity during a long sort, however, the purity of a sort may often need to be judged by standards that are different from those used to drive the sort decisions of the cytometer. Readers should look at the sorted cells in the right column of Figure 9.4 and guess at what factors may have contributed to less than perfect purity. In any case, a sort operator will ascertain the characteristics of the cells that are desired, set a sort gate that may be considerably tighter than those desired characteristics, and then check the success of the sort by comparing the sorted cells with those that were desired. Depending on the sophistication of the scientist commissioning the sort, all of this may be explicit or may just be part of the secret communication between the operator and his or her instrument.

At the end of the day, a flow sort needs to be characterized not only by the *purity* of the sorted cells according to the desired criteria but also by the *efficiency* of the sorting of the selected cells from the original mixed suspension (with low loss) and by the *time* taken to obtain the required number of sorted cells. In most cytometers, purity is good (understanding that apparent failures in purity may have legitimate reasons, as described above). Purity is well protected by electronic fail-safe controls that abort drop charging when cells are so close together at the analysis point that they may be enclosed in the same drop downstream (or in adjacent drops if two or three drops are sorted together). In this way, high cell flow rates will lead to lost cells and will therefore compromise the sorting efficiency but may not compromise purity until cells are so close together that they are coincident in the laser beam. As cell flow rates are increased by using more concentrated suspensions, more sort decisions may be aborted (leading to poorer and poorer efficiency), but the speed of sorting of the desired cells will increase until the cell flow rate gets so high and aborted sorts become so frequent that the actual speed of sorting of

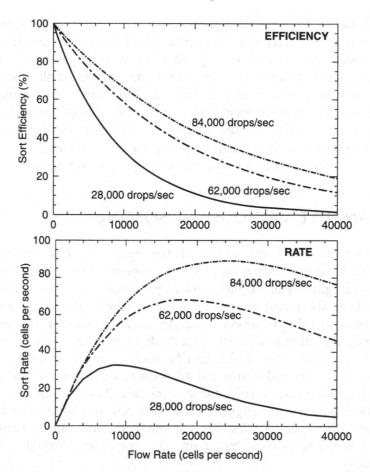

Fig. 9.5. Efficiency and speed of sorting are affected by the flow rate of cells. At high flow rates, more desired cells are lost, but the speed of collecting these desired cells increases until the loss of efficiency becomes greater than the increase in speed. High-speed sorting, with more drops per second, increases the efficiency and decreases the time required to obtain the desired number of cells. The model from which these graphs were generated was derived by Robert Hoffman: for these data, a three-drop sort envelope was used, 1% of the cells were "sorted," and the electronic dead time was set at 6 µs. If one drop is sorted with each sort decision (instead of three), the theoretical efficiency of the sorting improves considerably (as does the rate of collecting the sorted cells).

the desired cells starts to drop off (Fig. 9.5). This figure also indicates how beneficial high drop generation frequencies (high-speed sorting) can be for both the rate of sorted cell collection and also the efficiency of the sort.

Whether at high or low speed, every sort needs to be optimized according to the particular requirements of the experiment in question. Speed, purity, and sorting efficiency interact and are not, all three, optimizable under the same conditions. For example, a sort for a minor subpopulation of cells from a large, easily obtainable cell suspension may require optimization of speed, without concern about low sorting efficiency. On the other hand, a requirement for as many specific cells as possible from among a scarce mixed suspension will need to maximize recovery without much concern for speed.

ALTERNATIVE METHODS FOR SORTING

There are, of course, other methods for separating cells. Traditional methods have been based on physical properties (e.g., density gradient centrifugation) or on biochemical properties (e.g., the adherence of T cells to sheep erythrocytes or of monocytes to plastic or the resistance of white cells [as opposed to erythrocytes] to lysis by ammonium chloride). Additional separation methods involve the use of complement-fixing antibodies, which can be used to stain unwanted cells; complement lysis will enrich the suspension for the unstained cells. More recent methods involve the staining of cells with antibodies that have been bound with magnetic beads; the cells of choice, after being stained with the antibodies, become magnetic and can be removed from a mixed suspension by passing the suspension through a magnetic field.

All these methods are "bulk" or "batch" methods; that is, the amount of time that it takes to perform a separation is not related to the number of cells that are being separated. For example, if blood needs to be centrifuged for 20 minutes over a density gradient in order to separate monocytes and lymphocytes from neutrophils and erythrocytes, the separation will take 20 minutes whether the original volume of blood is 1 or 100 ml (assuming enough buckets in the centrifuge). Batch procedures are therefore ideally combined with flow sorting; they provide an initial rapid pre-separation that can greatly decrease the amount of time required for the final flow separation (refer back to Table 9.1).

By way of a practical example, consider CD5-positive B lymphocytes. Because CD5-positive B lymphocytes may be 10% of all B lymphocytes, B lymphocytes are perhaps 10% of all lymphocytes, and lymphocytes could be 50% of all mononuclear white cells, if we want to sort out CD5-positive B cells (0.5%) from a mononuclear cell preparation, they will be flowing through our system at a rate of 36,000 per hour if our total particle flow rate is limited to about 2000 per second. It would therefore take 28 h to collect 1 million CD5-positive B cells. However, if we remove all nonlymphocytes by adherence to plastic and then remove all T lymphocytes by rosetting with sheep erythrocytes (both techniques are rapid batch processes that might take an immunologist about an hour to perform), we can then send pure B lymphocytes through the cytometer. The desired CD5-positive particles (now 10% of the total) will therefore be flowing at a rate of 720,000 per hour. As a result, the time it will take to get 1 million desired cells will go from 28 hours to 1.4 hours. This has obvious benefits both in terms of cost, if you are paying for cytometer time, and the general health and viability of the sorted cells at the end of the procedure.

Batch procedures can be useful as pre-flow enrichment of the cells of interest. They can also be used as alternatives to flow sorting if the cells of interest can be appropriately selected. Flow sorting, although slow, is generally the only alternative if multiple parameters are required to define the cells of interest, if antigen density on the cell surface is low, or if forward or side scattered light is part of the definition of the desired cells.

In recent years, alternative flow cytometric methods for sorting have been developed. Instead of sorting cells by applying a charge to the drops generated by a vibrating stream, one of these methods involves sorting cells according to their flow signals by "grabbing" the cells of interest into a catcher tube that moves into the center of the stream when a desired cell flows by. Another flow sorting technology involves the use of a piston that applies pressure to one arm of a Y-shaped channel downstream from the laser intercept. Cells normally flow down the waste arm of the "Y." By application of pressure after a desired cell flows through the analysis point, selected cells are diverted from the waste arm to the sorting arm of the "Y." While these are cheaper alternatives than traditional, droplet-based, electronic sorting, speed limitations and the large volume of sheath fluid that dilutes the sorted cells make these methods primarily suitable for collection of small numbers of cells.

Another method for flow sorting is currently under development. It permits extremely rapid selection of cells at speeds beyond those possible even under very high-speed traditional (drop deflection) conditions. In drop deflection sorting, the sort rate is restricted by the rate of drop formation. This alternative method could be called *reverse sorting* or *optical zapping* and does not involve drop formation. Cells or chromosomes are stained with a phototoxic dye as well as with antibodies or a DNA stain. The particles are sent through a more or less conventional cytometer. Downstream from the analysis point, a cell or chromosome that does *not* possess the light scatter or fluorescent properties of interest is zapped by a second (killer) laser that activates the phototoxic dye. All particles end up in the collection vessel, but only those that have not been zapped survive. The rate of selection/destruction is limited, therefore, only by the rate at which the zapping laser can be deflected away from the stream. In proof-of-principle experiments, it appears that sorting rates may be increased 5-fold over those of the high-speed sorters and 25-fold over conventional sorters.

THE CONDITION OF CELLS AFTER SORTING

This brings us to a brief discussion of the condition of cells after they have been sorted. Depending on the purpose for which the sorted cells will be used, there may be different requirements for sterility and viability. If you are sorting cells for subsequent polymerase chain reaction amplification, for example, then neither sterility nor viability is important. On the other hand, if you are using a flow sorter because you want to clone cells expressing a certain characteristic, then both sterility and viability are necessary. Sterile sorting is possible (with care and a fair amount of 70% ethanol used to sterilize the flow lines). Sorted cells can remain sterile through the sort procedure and can subsequently be cultured for functional analysis or cloning. Given that cells in a sorting cytometer have been subjected to acceleration, decompression, illumination, vibration, enclosure, charging, and deflection toward high voltage plates, it seems reasonable to expect some trauma and, therefore, to coddle the cells as quickly as possible after the sort if viability is required. If you remember that the cells move down the stream in a core that is just a small percentage of the total volume of the sheath stream, you will realize that each drop is mainly sheath fluid and that, following drop deflection, the cells are deposited in medium that is partly sample medium, but mainly sheath fluid. Therefore, it is recommended to sort the cells into tubes that already contain a volume of appropriate culture medium. Some cells lose viability after sorting; many cells seem to be remarkably oblivious to the process.

It is also important to remember that most sorting procedures involve staining of the cells with antibodies conjugated to fluorochromes so that the cells can be sorted according to whether or not they are fluorescent. There is always the possibility that the binding of antibodies to the cell surface might by itself affect the function of the selected cells. In other words, the sorted cells that are low in fluorescence intensity may function differently from the cells that are bright not because they are essentially different but simply because the staining itself has blocked or tweaked surface receptors. By now you should be able to design a control for this potentially confusing phenomenon. If the staining is functionally neutral, stained cells (without any sorting at all) should function identically to unstained cells with regard to the assay of interest. In addition, it is important to be sure that the sorting procedure does not affect cell function. The control for this is to compare unsorted cells with cells that have gone through the sorter with large sort gates that do not actually exclude any cells. The sorting procedure should not change the functional ability of the cells if the proportion of different kinds of cells is not changed in this control sort. If the functional ability of the cells is changed either by sorting or by staining, then this needs to be considered in the interpretation of the subsequent functional analysis of the sorted populations.

FURTHER READING

Practical issues in sorting are discussed well in Chapter 4 of **Ormerod** and in Section 7 of **Diamond and DeMaggio**.

Theoretical principles of droplet generation sorting are discussed in Chapter 8 in **Melamed**, in Chapter 6 of **Watson**, and in Chapter 3 of **Van Dilla**.

High speed sorting is described in Section 1.7 of **Current Protocols in Cytometry**.

For discussion of chromosome sorting by optical zapping, see Roslaniec MC, Reynolds RJ, Martin JC, et al. (1996). Advances in flow cytogenetics: Progress in the development of a high speed optical chromosome sorter based on photochemical adduct formation between psoralens and chromosomal DNA. NATO Advanced Studies Series, Flow and Image Cytometry 95:104–114.

10

Disease and Diagnosis:
The Clinical Laboratory

Early in this book I stated that one of the areas in which cytometry
has had great impact is in clinical diagnosis. Research work, beginning
with Kamentsky's foray into cytological screening, but particularly
during the last 20 years, has made it clear that the information that
can be obtained by flow cytometry could be of clear use to clinicians.
However, as long as flow cytometers maintained their image of cumbersome and fiddly instruments, difficult to standardize and requiring
constant attention from a team of unconventional but devoted hackers
(see Figs. 1.3 and 1.4 for an understanding of the origins of this image),
there was little chance that flow cytometry would become integrated
into routine hospital laboratories.

The direct cause of the rapid introduction of flow technology into
many hospitals was the commercial development and marketing
of "black box" or "benchtop" flow cytometers (in imitation of the
pattern set by blood counters, which were, by the 1980s, standard
equipment in hematology laboratories). Installation of these benchtop flow instruments involves placing them on a laboratory bench
and plugging them in. They are optically stable and therefore can, in
principle, be maintained with relatively little human intervention;
they can produce clinical printouts that can be read without any requirement for knowledge of the intricacies of flow data analysis; and
they can, we are told, be run by anyone with the ability to push a key
on a computer keyboard (Fig. 10.1). It was only with the development and marketing of these friendly machines (starting in the mid to
late 1980s) that provision of flow cytometric information for routine
diagnosis became a practical proposition.

Fig. 10.1. Two opposing fantasies of what flow cytometry is all about. Drawings by Ben Givan.

As a result of the introduction of cytometers into the hospital setting, three aspects of clinical practice have led to some general reassessment of the nature of flow analysis. First, clinical laboratories are, because of the import of their results, overwhelmingly concerned with so-called quality control. This concern has forced all cyto-metrists to become more aware of the standardization and calibration of their instruments. Neither standardization nor calibration comes naturally to a flow cytometer. In response to this clinical requirement, beads, fixed cells, and mock cells have been developed to help in assessing the stability of conditions from day to day. Instruments can be set up from stored computer information so that machine parameters are constant from run to run. Furthermore, the analysis of data can be automated so that clinical information can be derived and reported quickly and in standard format. In addition, quality control schemes of various sorts have resulted in samples flying around the world; reports appear in the literature documenting the factors that lead to variation in results obtained from different laboratories and with different operators. Quality control in flow cytometry is still far from secure, but progress is being made toward producing technical recommendations (both for DNA and for leukocyte surface marker analysis), toward providing schemes for accrediting personnel and laboratories, and toward monitoring the performance of laboratories —with comparisons between laboratories and within any one laboratory over the course of time.

The second aspect of clinical practice that has led to a reassessment of the nature of flow cytometry is the occasional clinical requirement for "rare-event analysis." Methods have been developed, particularly with the use of multiparameter gating, to lower background noise in order to provide increased sensitivity for detection of rare cells. In the clinic, this increased sensitivity translates, for example, into earlier diagnosis of relapse in leukemia, more sensitive detection of fetal– maternal hemorrhage, and better ability to screen leukocyte-reduced blood transfusion products for residual white blood cells. Outside the clinic, these methods for rare-event detection have begun to stretch the limits of research applications as well.

A third aspect of clinical practice that has led to modifications in flow technology has been the requirement for safety in the handling of potentially infectious specimens. The fluidic specifications of instruments have been modified with attention to the control of aerosols that might occur around the stream, the prevention of leaks around the sample manifold, and the collection of the waste fluid after it has been analyzed. However, a more effective means of minimizing biological hazards has been the development of techniques for killing and fixing specimens in such a way that cells, viruses, and bacteria are no longer viable, but the scatter and fluorescent properties of the cells of interest are relatively unchanged. Formaldehyde (1.0%) is the fixative of choice in most flow laboratories. It is used after cells have been stained. As well as reducing the infectivity of the hepatitis B and AIDS (HIV) viruses, it fixes leukocytes in such a way that their scatter characteristics are virtually unaffected. Moreover, the fluorescence intensity of their surface stain remains essentially stable (albeit with slightly raised control autofluorescence) for several days or more. Therefore, routine fixation of biological specimens has not only increased the safety of flow procedures but has also made it possible to ship specimens around the world and to store specimens within a laboratory for convenient structuring of access to the cytometer (a euphemism for not working in the middle of the night). It should be said, however, that anyone working with material of known biological hazard needs to check fixation procedures to confirm the reduction of infectivity. Even after fixation, anyone working with any biological material at all should use standard precautions for control because any particular fixation procedure might not be effective against unknown or undocumented hazards. Furthermore, whatever fixation procedure is used should be checked with individual cell preparations and staining protocols to confirm the stability of cytometric parameters over the required period of time.

THE HEMATOLOGY LABORATORY

The flow cytometer has, for several reasons, found a very natural home in the hospital hematology laboratory. In the first place, because of the virtually universal use of automated instrumentation for enumerating erythrocytes and leukocytes, people working in hematology laboratories are quite relaxed about the idea of cells flowing in one end of an instrument and numbers coming out the other end. As I have said, both historically and technologically there is a close relationship between flow cytometers and automated hematology counters. The second reason is the obvious fact that blood exists as a suspension of individual cells—so that red and white cells do not need to be disaggregated before flow analysis. The third reason for the hematologist's ease with flow technology is that hematology/ immunology laboratories were among the first to make routine use of fluorescently tagged monoclonal antibodies. For many years, microscopists have been using panels of monoclonal antibodies for identifying various subpopulations of lymphocytes that are suggestive or diagnostic of various disease conditions. In particular, because the leukemias and lymphomas represent a group of diseases that involve the uncontrolled clonal proliferation of particular groups of leukocytes, various forms of these malignancies can be identified and classified according to the phenotype of the increased number of white cells found in biopsy material or in the patient's peripheral circulation.

Leukemia/Lymphoma Phenotyping

Studies of the staining of surface proteins (CD antigens) to determine the phenotypes of the abnormal cells in patients with various leukemias or lymphomas have been useful, revealing much about ways to classify the diseases but also increasing our knowledge of normal immune cell development. In general, immune cells gain and lose various surface proteins in the course of their normal development in the bone marrow until they become the cells (with mature phenotype and function) that are released from the marrow into the peripheral circulation (Fig. 10.2). Leukemias and lymphomas often involve a block in this development so that cells with immature phenotypes appear in great numbers in the periphery. Alternatively, certain forms of disease may involve rapid expansion of a clone of mature normal cells so that one type of cell predominates in the circulation over all other normal subpopulations. In other cases, cells seem to express abnormal combinations of CD antigens, a condition referred to as *lineage infidelity* or as *lineage promiscuity* depending on one's interpretation of the data. These variations in surface protein expression make classification of leukemias and lymphomas a suitable application for flow cytometry, albeit a very complex task. Correlation between classification and predicted clinical course is therefore correspondingly difficult.

Once the phenotype of a blood cell malignancy in a particular patient is known, the cytometrist can use the multiparameter flow description of that phenotype to define the malignant clone and to look for the absence of these cells in order to diagnose remission after

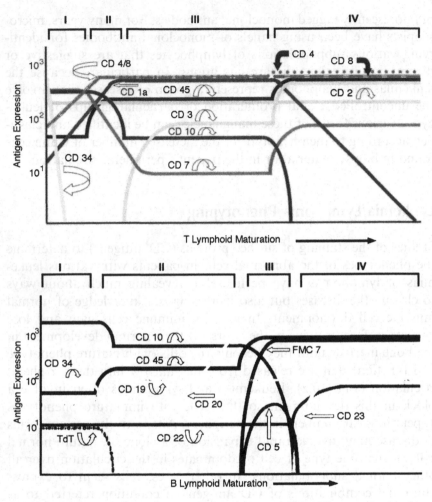

Fig. 10.2. Surface antigen changes during hematopoiesis. The upper plot is of T-lymphocyte maturation and the lower plot of B cells. From Loken and Wells (2000).

treatment. In addition, the flow phenotype of the malignancy can be used to look for the emergence of small numbers of these cells should relapse occur. This, of course, assumes that the relapse phenotype is identical to the abnormal clone defined at the initial diagnosis (Fig. 10.3).

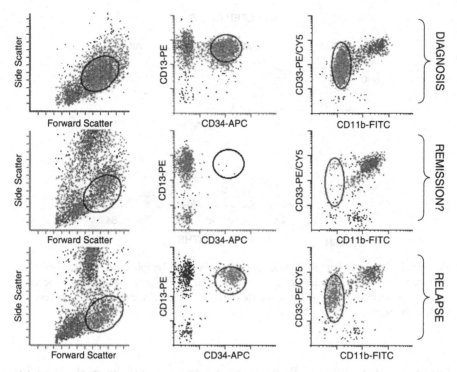

Fig. 10.3. Clusters of cells indicating an abnormal (acute myeloid leukemia) leukemic population, at diagnosis, at remission, and after relapse. The leukemic phenotype is CD13, CD34, and CD33 positive and CD11b negative. Courtesy of Carleton Stewart.

HIV/AIDS

Another condition that involves analysis of peripheral blood leukocytes is AIDS. Early in the natural history of the disease (or at least in the natural history of immunologists' awareness of the disease), it was discovered that one subpopulation of T lymphocytes in particular was destroyed by the HIV virus; the cells destroyed are those that possess the CD4 protein on their surface. It is this CD4 protein that appears to be a receptor involved in virus targeting. Therefore, much of the diagnosis and staging of AIDS involves the enumeration of CD4-positive cells in the peripheral blood (Fig. 10.4).

These techniques—for counting CD4-positive cells in connection with AIDS diagnosis and for phenotyping various populations of

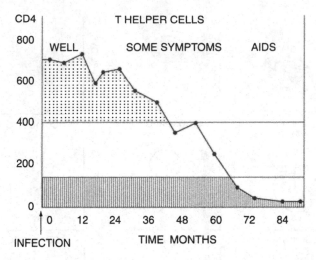

Fig. 10.4. The number of T-helper (CD4-positive) lymphocytes ($\times 10^3/cm^3$) in peripheral blood of a patient after infection with HIV. Cells <400 are significantly below the normal range; <100 indicates severe risk of clinical AIDS. Courtesy of Léonie Walker.

leukocytes for leukemia/lymphoma diagnosis—can be performed in hematology laboratories by the staining of cells with fluorochrome-conjugated monoclonal antibodies followed by the visual identification of different types of white cells and the counting of the fluorescent versus unstained cells under the microscope. Although the microscope has certain very definite advantages over the flow cytometer, two advantages it does not have are those of speed and statistical reliability. Particularly as a result of their work load from the growing number of AIDS patients, hematologists' need for a way to count statistically reliable numbers of cells from large numbers of patients became increasingly urgent. Flow cytometry was the obvious answer to this need.

Erythrocytes and Platelets

Although the initial perceived need in the hematology laboratory for a flow cytometer was to aid in the rapid processing of samples from leukemic and HIV-positive patients, the presence of the cytometer has stimulated thought about new hematological applications. Although not yet in routine use for analyzing erythrocytes, flow cytometers have been shown to be useful for looking at red-cell–bound immunoglobulin (as a result of autoimmune disease, sickle cell anemia, and thalassemia): The bound immunoglobulin on erythrocytes is detected by the use of fluorescent antibodies against human immunoglobulin. The staining of red cells for RNA content with a dye called *thiazole orange* has made possible the use of flow cytometry to count the reticulocytes (immature erythrocytes) present in blood samples from anemic patients.

Hematologists have also extended the use of flow analysis to platelets—those particles with low forward scatter that usually are ignored in flow cytometric applications because they fall well below the standard forward scatter threshold. Immunoglobulin bound to platelets can be measured with antibodies against human immunoglobulin (as in the detection of immunoglobulin on erythrocytes). Platelet-associated immunoglobulin (resulting from autoimmune disease) is determined by analyzing the patient's platelets in their natural state. In another clinical situation, antibodies in the serum with specificities for "foreign" platelets may develop after pregnancy or transfusions; these can be monitored by flow cytometry if a patient's serum is incubated with a donor's platelets. Recently, the activation state of platelets has also been analyzed. Hyperactive platelets express P-selectin (CD62) on their surface; they have been primed to facilitate coagulation. Although not yet applied in routine diagnosis, research indicates that expression of CD62 may provide prognostic information in myocardial infarction and cardiopulmonary bypass surgery.

HLA-B27 Typing

Human cells express a series of proteins on their surface that are involved in self-recognition and in antigen presentation to effector cells for the stimulation of immune reactions. These are called *major histocompatibility* (MHC) antigens, and each occurs in a variety of allelic forms. People, therefore, differ from each other in their MHC genotypes. As well as affecting transplant compatibility, certain genotypes have been implicated in predisposition to autoimmune disease.

In particular, HLA-B27 (the 27th haplotype at the B locus of the human leucocyte antigens), has been associated with arthritic conditions (ankylosing spondylitis, juvenile rheumatoid arthritis, and Reiter's syndrome). While only about 20% of people with the HLA-B27 genotype have ankylosing spondylitis, for example, a very high percentage (perhaps 90%) of people with this disease are HLA-B27 positive (compared with 4–8% in the general population). Antibodies against the HLA-B27 protein are available. Leukocytes can be incubated with fluorochrome-conjugated antibodies; positive staining indicates expression of the HLA-B27 protein. This assay can be performed with a microscope (using either fluorescent antibodies or complement-fixing antibodies that kill cells). Recently, increasing numbers of hematology laboratories have begun to use flow cytometers for this assay.

CD34 Stem Cells

After a cancer patient has been treated with irradiation to destroy malignant cells, normal blood cells will also be depleted and the patient may require transplantation to regenerate these cells. Hematopoietic stem cells (classified by possession of the CD34 antigen on their surface) are immature, undifferentiated cells that have the ability to differentiate into all classes of blood cells. They exist, normally, in the bone marrow; detectable numbers can be found in the peripheral blood if the patient has been treated with blood cell growth factors. If enough of these stem cells can be harvested from the patient before irradiation, transplantation of these cells back into the patient after irradiation can replenish the immune system. The number of these CD34-positive cells given back after irradiation needs, therefore, to be assayed in order to ensure that the stem cell transplant is sufficient to return the patient to normal immune competency. Because these stem cells are uniquely characterized by the CD34 antigen, flow cytometry can be used to confirm the presence of sufficient stem cells in a cell suspension obtained from the patient before irradiation (Fig. 10.5). Because the actual number of these stem cells (rather than just their proportion in a 10,000 cell data file) is important, the volume flowing through the flow cytometer can be calibrated by the addition of known numbers of beads to the sample.

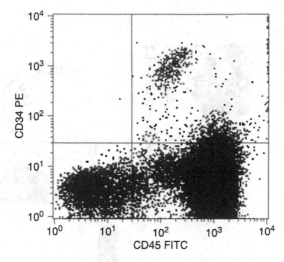

Fig. 10.5. Stem cells can be assayed by the use of antibodies against the CD34 and CD45 antigens. From R Sutherland as published in Gee and Lamb (2000).

Fetal Hemoglobin

Detection of fetal erythrocytes in the maternal circulation is important in diagnosing a cross-placental fetal–maternal hemorrhage that might have resulted from trauma with suspected maternal injury. In addition, in the case of Rh-blood group incompatibility between fetus and mother, rapid detection of fetal cells in the maternal circulation can allow early intervention and the prevention of Rh-incompatibility disease (erythroblastosis fetalis) in the newborn. Although detection of these fetal cells is difficult because their proportion in the maternal circulation is low (less than 0.1% in a normal pregnancy; perhaps 0.6% in a pregnancy that needs therapeutic intervention), they can be assayed by the use of antibodies against the fetal form of hemoglobin (HbF). This procedure is typical of flow protocols concerned with rare-event analysis (Fig. 10.6). Defining the cells of interest (in this case the HbF-positive cells) by multiple parameters (forward scatter, side scatter, autofluorescence, HbF) reduces the likelihood of false positivity resulting from background noise.

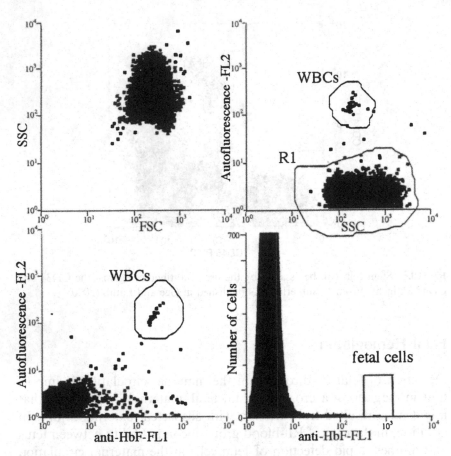

Fig. 10.6. Detection of rare fetal erythrocytes in the maternal circulation. The fetal erythrocytes are HbF positive and low in autofluorescence. The cells in R1 are displayed and enumerated in the bottom right histogram. From Davis (1998).

THE PATHOLOGY LABORATORY

Tumor Ploidy

In the chapter about DNA analysis, I mentioned the large amount of work generated for flow laboratories as a result of publication of the Hedley technique for analyzing the DNA content of paraffin-embedded pathology specimens. After the first headlong rush of publications correlating DNA ploidy with long-term prognosis in various types of cancer, the field has now settled down a bit. As I have indicated, it would probably be fair to say that most (but not all) studies on disaggregated solid tumors (fresh, frozen, or fixed) have shown some kind of correlation between abnormal DNA content and unfavorable long-term prognosis.

There is considerable debate in the literature about whether flow cytometric analysis of ploidy gives any additional prognostic information that is independent of other known prognostic indicators, but most studies have indicated that it does. It appears to be useful, for example, in indicating, from all breast cancer patients without lymph node involvement, a group of women who are at risk for recurrence of disease (Fig. 10.7). Because any pathological material will be a mixture of both normal and abnormal cells, there is much current effort expended on ways of selecting from among the cells of a disaggregated specimen those particular cells that are from the tumor itself. If those tumor cells can be stained selectively (as with a fluorescein-conjugated anti-cytokeratin antibody for epithelial cell-derived tumors within nonepithelial tissue) and then the DNA con-

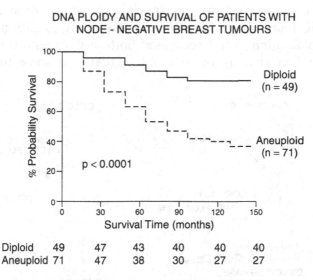

DNA PLOIDY AND SURVIVAL OF PATIENTS WITH
NODE - NEGATIVE BREAST TUMOURS

Diploid	49	47	43	40	40	40
Aneuploid	71	47	38	30	27	27

Fig. 10.7. Survival curves for breast cancer patients without lymph node involvement. Those with DNA aneuploid tumors (as diagnosed by propidium iodide flow cytometry of 10-year-old paraffin-embedded material) survived significantly less long than those with DNA diploid tumors. Graph from Yuan et al. (1991).

tent of the fluorescein-stained cells determined, we may have a dual color flow method of much improved sensitivity for detecting aneuploid cells from within a mixture of both malignant and normal components.

S-Phase Fraction

There is also evidence, in solid tumors as well as in the bone marrow plasma cells from patients with multiple myeloma, that determination of the proliferative potential of malignant cells is a better indicator of poor prognosis than simply the determination of the presence or absence of aneuploid cells. As discussed in the chapter on DNA, proliferation (the proportion of cells in S phase) can be assessed by using mathematical algorithms to analyze the shape of the DNA histogram. However, the presence of more than one cell type, with the G2/M peak from the diploid cell line overlapping the S phase of the aneuploid cells, makes mathematical analysis even more complex than with distributions resulting from single euploid cells or clones (Fig. 10.8). Nevertheless, correlations between high S-phase fraction and poor clinical prognosis appear relatively strong. Bromodeoxyuridine (BrdU) has been used in the research setting to give information about proliferation; it has been used both in vitro with fresh excised tumors and in vivo by infusion into patients at some time before

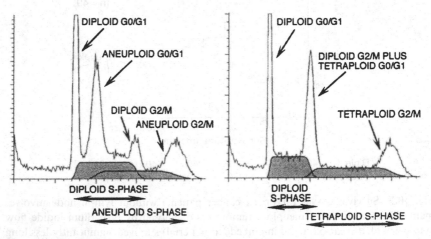

Fig. 10.8. Analysis of mixed diploid/aneuploid cells for the S-phase fraction of the aneuploid population.

excision of the tumor. Estimates of tumor doubling time, based on the rate of movement of BrdU-pulsed cells through the cell cycle and on the percentage of cells in the tumor that are dividing, have been shown to correlate with aggressiveness of the malignancy. There may in the future be more use of this technique for assaying malignancies for sensitivity to various cytotoxic drugs.

Ploidy and S-phase fraction determinations are, in most cases, the only flow analyses that have achieved much routine application in the field of solid tissue oncology. At the present time, it would appear that flow cytometry has not settled quite so comfortably into the pathology laboratory as it has into the hematology setting. This may, I suspect, come from the crucial difference between hematology and pathology—and it is a difference that may remind us of one very definite limitation of flow analysis. Pathological specimens come from solid tissue; to analyze them with a flow cytometer, the solid tissue must be disaggregated in some way. This disaggregation process (either mechanical, with detergent, or with enzymes) not only has a good chance of compromising the integrity of some or many of the cells we want to analyze, but flow cytometry will also, necessarily, lose any information that is contained in the structural orientation or pattern of the original cells and tissue, as viewed under the microscope.

A response to this dilemma has come, recently, in the development of laser-scanning cytometry (the LSC by CompuCyte). The LSC uses a laser to scan cells or tissues on slides and acquires data that can be presented much as any flow cytometric fluorescence or light scatter information (in histograms or two-color plots). However, the LSC also acquires extra parameters that record the x/y position of each cell on the slide. In this way, cells that show aberrant fluorescence or scatter characteristics can be examined individually under the LSC's microscope to provide high-resolution images and confirm malignant or otherwise interesting phenotype. In many ways, the LSC bridges the technologies of flow cytometry and traditional microscopic pathology.

SOLID ORGAN TRANSPLANTATION

It has been known for many years that transplant recipients may possess serum antibodies that will react with and destroy cells from a transplanted organ. If these so-called cytotoxic antibodies are present at the time of transplantation, then an immediate and violent rejection crisis will occur (hyperacute rejection), destroying the grafted organ and threatening the life of the recipient. Therefore, the serum from a potential recipient is assayed before surgery for cytotoxic antibodies by mixing it with lymphocytes from the organ donor (in the presence of rabbit complement). Traditionally, the cytotoxicity of the serum in this cross-match assay is scored by counting dead lymphocytes (cells taking up propidium iodide or trypan blue) under the microscope. Addition of anti-human globulin can enhance the sensitivity of this assay for bound immunoglobulin.

In the early 1980s, MR Garovoy and his group in California suggested that this cross-match assay might not be sensitive enough. Some of the cases of slower rejection might result from the presence in the organ recipient of levels of preformed antibodies that were too low to be detected by the traditional cytotoxic assay. A flow cytometric cross-match assay for these preformed antibodies directed against donor lymphocytes was developed by Garovoy and subsequently modified by Talbot in Newcastle and by workers at other centers to make use of two-color fluorescence. The assay involves the incubation of the organ donor's lymphocytes with serum from potential recipients followed by the staining of the lymphocytes with a phycoerythrin-conjugated monoclonal antibody against either T cells or B cells and a fluorescein-conjugated antibody against human immunoglobulin (Fig. 10.9). In that way, it can be determined whether the recipient has antibodies that coat the T lymphocytes or the B lymphocytes of the donor (Fig. 10.10). This two-color technique (or three-color, if T- and B-cell antibodies are used simultaneously) can in fact serve to remind us of a general approach to the design of flow cytometric assays: One color is used to pinpoint the cells of interest (in this case, either T cells or B cells), and the second color is used to ask some question about those cells (in the cross-match, we are asking whether those T or B cells have become coated with immunoglobulin from the recipient's serum).

Evidence over the past 15 years has indicated that this assay for lymphocytes with bound antibody is indeed more sensitive in detecting antibodies directed against donor cells than is the traditional (microscope) assay for lymphocyte death. The flow assay also defines a group of recipient–donor pairs who are at risk for severe rejection crises and possibly, but not necessarily, for loss of the grafted organ. At the present time, donated organs are in short supply. One transplant

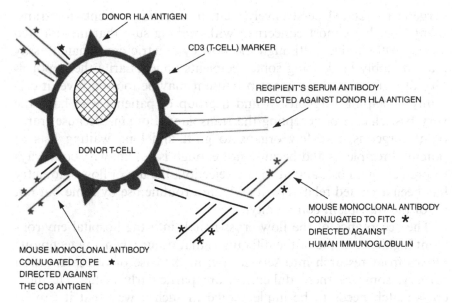

Fig. 10.9. Protocol for the dual-color flow cytometric assay for testing recipient serum for the presence of antibodies to donor cells before organ transplantation.

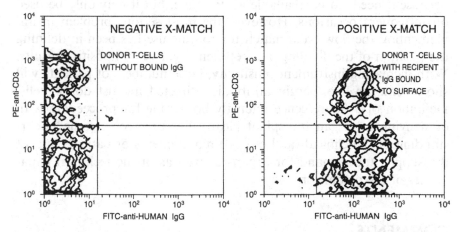

Fig. 10.10. Contour plots indicating a cross-match assay for immunoglobulin from the recipient bound to the surface of the donor's T cells. Data from the Tissue-Typing Laboratory, Royal Victoria Infirmary, Newcastle upon Tyne.

surgeon has stated persuasively that, in selecting patients for transplantation, he is most concerned with making sure that the donated
kidney will survive. Although the flow cytometric cross-match assay
may arguably be denying some recipients unnecessarily the opportunity of a successful transplant because it may be too sensitive, it certainly is helping surgeons to find a group of patients who have the
very best chance of accepting the transplanted organ. Because transplant surgeons have few organs to graft and long waiting lists of
potential recipients and because not enough is yet known about ways
to predict, diagnose, and abrogate rejection, the use of flow cytometry
has been accepted relatively quickly into routine use by some but not
all of the transplant community.

The acceptance of the flow cross-match into the hospital environment highlights some of the difficulties that occur when any technique
moves from research into service use; in the case of transplantation
surgery, some of these difficulties are particularly acute. The flow
cross-match needs to be implemented in such a way that it can be
performed by people who may be working in the middle of the night
under pressure of time and without specialist support staff available
for easy consultation. Under such conditions, automated software
and stable instrumentation are especially important. In addition, flow
cytometers are relatively expensive instruments; tissue-typing laboratories are concerned with ways to justify the expense of a cytometer
because it needs to be available all the time, but it may only be used
on infrequent occasions. However, the most serious problem in im-
plementing the flow cross-match for routine use has been in defining
the position of the dividing line between negative and positive results.
With increasing instrument sensitivity, large numbers of patients will
show some level of serum antibodies directed against donor cells.
Definition of a fluorescence intensity borderline for proceeding with
or denying an organ transplant clearly has to involve some understanding of the clinical (and ethical) requirements of patients and of
the supply and demand for organs as well as of the technical capabilities of the cytometer.

COMMENTS

Rational decisions still need to be made about the fields in which flow
cytometry can make a positive contribution to patient care, about the

methods available for quality control, and about the kind and depth of training required by staff. Good channels of communication also need to be opened between laboratory staff and the clinicians who are making diagnostic use of flow data, but who may not be aware of its limitations and conventions. Benchtop cytometers are optically stable and can be used and maintained with remarkably little human intervention. Software packages for those cytometers can apply algorithms for automated lymphocyte gating of blood preparations—the computer will draw a lymphocyte gate for you. The future may contain more sophisticated expert systems employing artificial intelligence in order to avoid the pitfalls associated with human judgement. How stable these instruments are in practice, whether automated computer algorithms make reasonable guesses with difficult samples, and whether an untrained scientist or clinician can draw reliable conclusions from automated print-outs of flow data are open questions. With flow cytometry, as in any field of science, we need to guard against conveying a false sense of objectivity to laymen not familiar with the subjectivity of a technique.

Kenneth Ault, the former president of the International Society for Analytical Cytology, summed up some of flow cytometry's clinical growing pains with the following statement:

> Flow cytometry is a technology that seems to stand at the threshold of "clinical relevance." Those of us who have been using this technology, and especially those who are manufacturing and selling the instruments and reagents, are frequently evaluating the status of clinical flow cytometry. Most of us have little doubt that this is going to be an important technology in clinical medicine for many years to come. However, from my point of view, and I think for many others, the movement into the clinic has been unexpectedly slow and painful.... In talking about this in the past I have frequently mentioned the "grey area" between research and clinical practice. I was recently reminded of a quotation from T.S. Eliot: "Between the idea and the reality falls the shadow." I believe that flow cytometry is currently traversing that shadow. KA Ault (1988). Cytometry (Suppl 3):2–6.

That paragraph was written for a talk given in 1986. In the first edition of this book (1992), I said that clinical cytometry was, at that time, in what we might call dappled sunlight. Fourteen years after Ault's talk, in 2000, Bruce Davis, the president of the Clinical Cytometry Society (Newsletter 5:1–2) said that he remains cautiously optimistic regarding the future of diagnostic cytometry in areas of both image and flow cytometry. Although the applications of lymphocyte subset counting have seemed to reach a plateau, new applications using cytometry techniques for disease diagnosis, monitoring, and prognostication continue to appear.

Flow cytometry appears fairly comfortable, now, in many hospital laboratories, and most cytometrists might agree with Davis's cautious optimism. It may be, however, that flow cytometry has not yet reached the full glare of total integration into the hospital community.

FURTHER READING

There are many books that cover, in detail, the aspects of clinical flow cytometry that I have mentioned in this chapter and many other aspects as well. Some of the books that I have found most useful are listed here.

Bauer KD, Duque RE, Shankey TV, eds. (1993). Clinical Flow Cytometry: Principles and Applications. Williams & Wilkins, Baltimore.

Keren DF, Hanson CA, Hurtubise PE, eds. (1994). Flow Cytometry and Clinical Diagnosis. ASCP Press, Chicago.

Laerum OD, Bjerknes R, eds. (1992). Flow Cytometry in Hematology. Academic Press, London.

Landay AL, Ault KA, Bauer KD, Rabinovitch PS, eds. (1993). Clinical Flow Cytometry. New York Academy of Sciences, New York.

Owens MA, Loken MR (1995). Flow Cytometry Principles for Clinical Laboratory Practice. Wiley-Liss, New York.

Riley ME, Mahin EJ, Ross MS, eds. (1993). Clinical Applications of Flow Cytometry. Igaku-Shoin, New York.

Stewart CC, Nicholson JKA, eds. (2000). Immunophenotyping. Wiley-Liss, New York.

11

Out of the Mainstream: Research Frontiers

In the previous chapters, I have discussed what may be thought of as mainstream applications of flow technology. Cells stained for surface and intracellular antigens and nuclei stained for DNA content together constitute a large majority of the particles that flow through the world's cytometers. As noted in the Preface, however, flow cytometry has continued to surprise everyone with its utility in unusual and unpredicted fields of endeavor. By the time this book appears in print, some new applications will almost certainly have progressed into the flow mainstream and other newer applications will have taken their place in the tributaries (is it time to kill this metaphor?).

At present, the large number of "tributary" applications (highly varied, often idiosyncratic, and sometimes unsung) precludes any attempt at making a chapter in an introductory book such as this into a complete catalogue. Therefore, I will attempt here only to hint at a few applications that seem to me to be important, or unusual, or to show promise (it has not escaped my attention that several of the applications that I said showed promise 10 years ago are still included in this chapter as holding potential ...). Because each reader will have his or her own personal goals for the use of flow cytometry according to his or her own interests, I think it is important here mainly to convey some feeling for the enormous variety and surprising potential that are found in laboratories using flow cytometry as a research tool.

FUNCTIONAL ASSAYS

Activation Markers

Functional analyses can be performed on the flow cytometer by assaying cells at different times during the progress of a stimulated or ongoing change. For example, although leukocytes express certain proteins on their surface that can be thought of as phenotypic markers (in other words, these markers are relatively constant in intensity and are expressed dependably by certain subgroups of cells), other proteins come and go on the cell surface, depending on the cell's functional/physiological/activation state. These surface proteins (the interleukin-2 receptor [CD25] or CD69 or CD154 on T cells are good examples for leukocytes) show characteristic time courses. Time of appearance after stimulation varies, and expression of some is long-lasting and of others is relatively transient (Fig. 11.1). In addi-

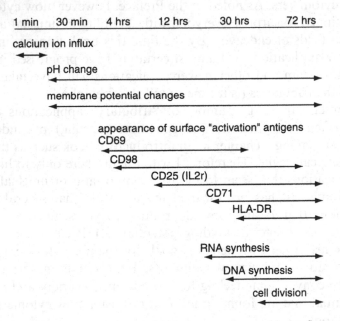

Fig. 11.1. Lymphocyte responses to stimulation, as measurable by flow cytometry. The sequence and duration of responses are approximate and will depend on the nature and strength of the trigger. Adapted from Shapiro (1995).

tion, there are internal proteins that are synthesized in response to cell activation: The cytokines discussed in the chapter on intracellular proteins are in this class, as are the cyclins discussed in the DNA chapter.

This pattern of progressive and/or transient expression raises certain problems for a flow cytometrist interested in knowing about the physiological state of a cell. Obviously, the time at which one assays the cells is critical, but it should also be clear that any protein that comes and goes will also, at certain times, be expressed at very low levels on the cell surface. Therefore it becomes essential to know something about the sensitivity of the flow cytometer. A cell may appear to be negative for a certain protein if it is assayed on a certain cytometer using a certain fluorochrome-conjugated antibody, but the protein might be detectable if a more sensitive cytometer or a brighter fluorochrome is used. The other issue with so-called activation markers is that the level of expression becomes the relevant read out (Fig. 11.2). We become interested not just in whether a cell is positive but in how positive it is (the median or mean intensity rather than the proportion of positive cells). Calibration of the fluorescence scale also becomes more important if we want to compare intensity values from day to day or from laboratory to laboratory.

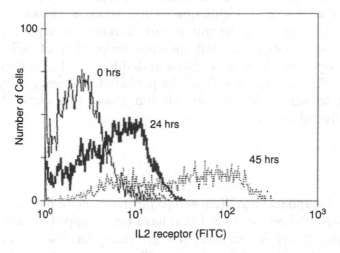

Fig. 11.2. The gradual increase in expression of interleukin-2 receptors on the surface of lymphocytes after phytohemagglutinin stimulation.

Precursor Frequencies

By way of a completely different type of functional analysis, we can look at the use of tracking dyes to label and follow cells. Cells can be stained with lipophilic fluorochromes or with fluorochromes that bind nonspecifically to all proteins. An example of the lipophilic fluorochromes are the so-called PKH dyes that insert themselves into the bilayer of cell membranes. The dye carboxyfluorescein succinimidyl ester (CFSE) binds to the free-amino groups on all cell proteins. These two types of dyes can be used to stain cells with considerable stability. The stained cells can then be injected into an animal and tracked to their homing location.

In a different use of this technique, the proliferation history of a cell can be followed. Because the dye is passed on from a parent to each of its daughter cells, a halving dilution of dye content occurs at each cell division (and a resulting halving of intensity is seen). In this way, according to their intensity, the cells in a suspension can be classified as to how many divisions they have undergone since their labeling at the beginning of the experiment. Those with half the original intensity have undergone one cell division; those with one eighth of the original intensity have undergone three cell divisions; and so forth. By analysis of the intensity histogram of a population of cells after several days incubation with a stimulus (knowing that a cell's intensity gives us the number of divisions undergone by that cell), back-calculation can tell us what proportion of cells in the original population were stimulated to divide (Fig. 11.3). This method is therefore a way to look at the precursor frequency of cells responding to various specific stimuli and a way to look at the past proliferative history of a given cell.

Kinetics

Whereas the applications discussed so far in this book have dealt with ways to describe cells or nuclei that have been stopped in their tracks for analysis at a given moment, flow cytometry has also been used to follow the physiological function of cells in true kinetic analysis. As an example of this kind of analysis, we can discuss the use of a flow

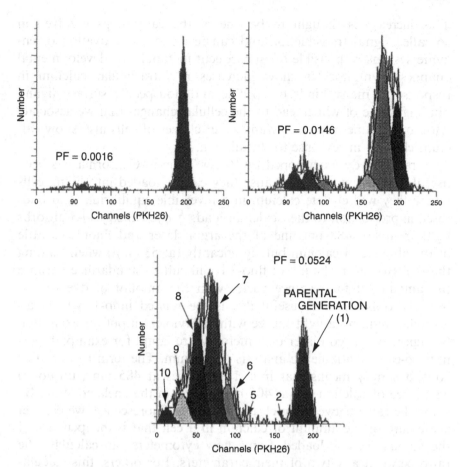

Fig. 11.3. The analysis of a histogram of PKH26 fluorescence to categorize cells according to their generation number (1–10) after 10 days culture. As cells divide, the PKH dye becomes diluted by half, with each halving of intensity marking cells as having gone through another division. The precursor frequencies (PF) of the dividing cells can be calculated. Three examples are shown with lymphocytes from donors with low, medium, and high responses to the tetanus toxoid antigen. Data from Jan Fisher and Mary Waugh.

cytometer to look at the rapid changes in calcium ion concentration that occur when cells become physiologically activated.

One of the early responses made by many types of cells to a stimulus is a rapid influx of calcium ions across the plasma membrane and a resulting increase in the cytoplasmic level of free ionic calcium.

This increase is thought to be one of the early steps involved in so-called signal transduction and can result in the activation of enzyme systems responsible for subsequent metabolic or developmental changes. Lymphocytes show increases in intracellular calcium in response to many kinds of specific and nonspecific surface ligand binding, some of which lead to the cellular changes that we associate with an immune response. Many other classes of cells also show calcium changes in response to stimulation.

A range of dyes developed by Roger Tsien in California has been useful in flow cytometry because they can be loaded into living cells where they will chelate calcium in a reversible equilibrium and fluoresce in proportion to their calcium load. A dye called *fluo-3* absorbs light from the 488 nm line of the argon laser and fluoresces little in the absence of calcium but significantly (at 530 nm) when binding the ion. However, the use of fluo-3 is difficult to standardize because the amount of fluorescence varies with the amount of dye loaded into the cells. A more useful dye is the related indo-1, which, although it requires a light source with ultraviolet output for excitation (a high-power argon laser or a mercury arc lamp, for example), permits so-called ratiometric analysis of calcium. The term *ratio* in this context simply means that indo-1 fluoresces at 485 nm (turquoise) when free of calcium but at 405 nm (violet) in the chelated state. By using the ratio between violet and turquoise fluorescence, we can get a measure of the amount of calcium in a cell that is independent of the amount of dye loaded. Some flow cytometers can calculate the ratio between any two of their parameters. For others, this calculation has to be done through post-acquisition software.

Figure 11.4 shows a time course of this ratio (related to internal calcium) as it changes in response to the stimulation of lymphocytes by a ligand bound to their surface receptors. Such a time course can be performed on a flow cytometer by loading cells with indo-1 in advance and then stimulating them with a ligand (in a calcium-containing medium) immediately before placing the sample tube on the sample manifold. Alternatively, the sample manifold of many research cytometers can be modified to permit rapid injection of reagents into the sample tube while the sample is being drawn through the system. In either case, measured cell parameters can then be acquired over the subsequent minutes and their 405:485 fluorescence ratio determined.

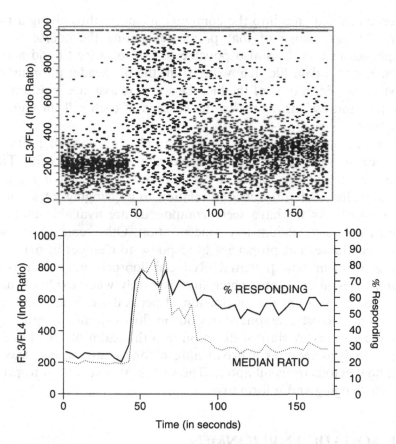

Fig. 11.4. The time course of a calcium response (measured by the indo-1 405/485 fluorescence ratio) induced in lymphocytes by the addition of a stimulus. The upper plot is a two-dimensional dot plot of the calcium ratio versus time. The lower plot shows data binned in 5 s intervals and processed to give both the median ratio and the percentage of cells above the base line as they change with time.

Of course, to talk about kinetic measurements, we need to bring in the parameter of time. Time is, in many ways, the hidden extra parameter of every flow system. Depending on the software being used, it will be more or less easy to access this time parameter for use in a kinetic profile. In the best case, each data file can have time as an extra parameter for each cell. We would be able to plot any other parameter(s) like turquoise and/or violet fluorescence against time for each cell acquired into the computer memory, thus giving a time course of the change of that parameter during the course of the sample acquisition. It is also possible to use software to add a time parameter to a data file. Knowing, from the file header information, the start and finish time of the file, the software assumes constant cell flow rate and assigns a time to the passage of each cell through the laser beam.

This calcium ion procedure is an example of a class of protocols that measure the kinetics of functional activity in living cells. They depend on the ability to find a compound able to enter living cells and then alter its fluorescence in relation to a changing intracellular environment. As we have seen, compounds are available that can be used to assay calcium ion concentration. Other compounds will alter their fluorescent properties in response to changes in pH or to changes in membrane potential. Reduced fluorochromes such as di-chlorofluorescin diacetate, which fluoresce only when oxidized, have been used to measure the production of peroxides during neutrophil activation. Almost everyone starts out in flow cytometry with work on static systems. A shift of direction into the realm of function and kinetic analysis requires a broadening of one's entrenched ideas of what flow cytometry is all about. The shift is almost certain to prove both challenging and informative.

THE AQUATIC ENVIRONMENT

Blood has been an ideal object for flow cytometric attention in large part because it occurs naturally as a mixed population of single cells. Lakes and oceans are also suspensions of mixed types of single cells. As such, they would appear to present an obvious target for flow analysis. Indeed, flow cytometers are now in place in many marine laboratories and on board sea-going vessels.

Aquatic single cell organisms with a size range of about 0.02–200 μm in diameter occur in nature in concentrations that range from about 10^2 to 10^7 per ml. They include viruses, bacteria, cyanobacteria (formerly known as *blue-green algae*), autotrophic phytoplankton (unicellular plants), and heterotrophic zooplankton (unicellular animals). The analysis of aquatic organisms by flow cytometry presents some characteristic features that may serve to highlight the issues that, to a greater or lesser extent, affect all flow analyses. In the first place, because of the presence of naturally occurring photosynthetic pigments, the phytoplankton are highly autofluorescent (recall that some of them contain phycoerythrin, peridinin-chlorophyll complexes, or allophycocyanin). This autofluorescence leads to high background intensity against which positive staining of low intensity may be difficult to detect. The autofluorescence is also variable and may depend on the environment or metabolic state of the cell. The autofluorescence can, however, be exploited and used to distinguish different classes of organisms and different metabolic states.

Another characteristic of the aquatic environment is that the abundance of organisms of different types is highly variable; aquatic scientists do not have the benchmarks of a fairly tight "normal range" that clinical scientists depend on. In addition, the abundance of very small cells in the aquatic environment presents a challenge in instrument tuning and sensitivity. Not all cytometers can distinguish the forward scatter signal of nano- or picoplankton from optical noise or from particulate matter in the sheath stream. Because of the great size heterogeneity of plankton, a cytometer for aquatic analysis must be able to cope with both small and large particles at the same time. A final problem is that the most common particles in aquatic samples are not living; they represent decaying organic matter, silica- or calcium-containing empty cell walls, and suspended sediment, which are all difficult for a flow cytometer to distinguish from living cells.

Despite these problems, flow cytometry has had some noted success in aquatic research, particularly in relation to studies on the phytoplankton. Because all phytoplankton possess chlorophyll, but only the cyanobacteria possess the phycobiliproteins, autofluorescence "signatures" from water samples, based on the chlorophyll (fluorescence >630 nm), phycoerythrin (fluorescence <590 nm), and forward scatter of particles, have been used to characterize the changes that occur in plankton at different depths or at different locations (Figs. 11.5 and 11.6).

Figure 11.7 shows an example of the way in which flow cytometric analysis can distinguish six different species of plankton in culture and define which of these species are favored by grazing marine scallops as a source of food. Results such as these have been used to

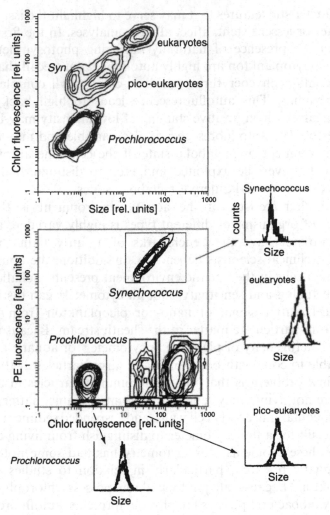

Fig. 11.5. The flow cytometric signature of a seawater sample taken at 90 m depth in the North Atlantic. Chlorophyll autofluorescence (>650 nm) has been plotted against side scatter ("size") and against phycoerythrin autofluorescence (530–590 nm). From Veldhuis and Kraay (2000).

suggest modifications in the menu supplied to scallops being farmed in aquaculture tanks.

Flow cytometry has also led to the notable discovery, reported by Sallie Chisholm in 1988, of the existence of a novel group of small,

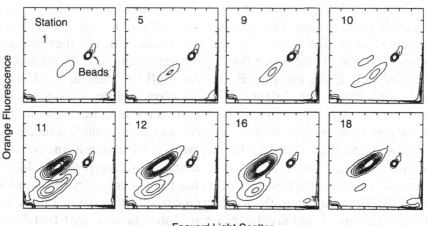

Fig. 11.6. Flow cytometric analysis of surface water from points at 1.5 mile intervals off shore from Cape Hatteras, North Carolina. Forward scatter and orange auto-fluorescence identify two *Synechococcus* populations with different phycoerythrin content. Beads were used to calibrate the number of cells present. From Chisholm et al. (1986).

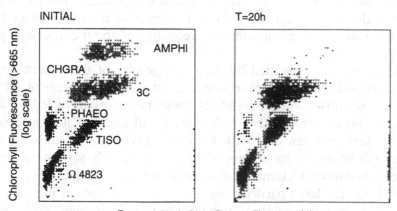

Fig. 11.7. Small scallops were placed in tanks with six species of phytoplankton that show distinctive flow cytometric signatures. After 20 h of grazing, it was apparent from the differences between the flow dot plots that the scallops had exhibited definite preference for two of the six species. Courtesy of Sandra Shumway.

prokaryotic phytoplankton; these free-living, marine prochlorophytes (*Prochlorococcus*) are between 0.6 and 0.8 μm in size but possess pigments more like those of eukaryotic plants than of other prokaryotes. The use of shipboard flow cytometry during cruises off southern California, the Panama Basin, the Gulf of Mexico, the Caribbean, and the North Atlantic between Woods Hole, Massachusetts and Dakar, Senegal (who said flow cytometry isn't fun?) has found these previously unknown prochlorophytes in remarkable abundance and indicated that they may be responsible for a significant portion of the global photosynthetic productivity of the deep ocean. More recently, Claude Courties (1994) has used flow cytometry on samples from the Thau lagoon in the Mediterranean off the coast of France to detect very small, eukaryotic picoplankton (about 1 μm) that have been called "the smallest eukaryotic organisms."

The CytoBuoy project from the European Community Marine Science and Technology Programme has been developing prototypes for a small flow cytometer (a 38 × 48 cm cylinder) that fits in ocean buoys. Current implementation of the CytoBuoy includes a cytometer inside a buoy, sampling just below the water surface. Forward scatter, side scatter, and orange fluorescence signals from particles are collected. It is projected that the CytoBuoy could take samples in the ocean at depths up to 500 m. Data transmission for continuous monitoring of ocean plankton could occur via short wave over a distance of up to 50 km. At a rate of about 20 bytes per second, this would give a sampling frequency of about one sample per hour.

The problems presented by the heterogeneity of the aquatic environment and the instrumental and conceptual developments made by aquatic scientists toward handling these problems have led to advances that can enrich the work done in all fields of flow analysis. Cytometers that can deal with the instability of the ocean environment will be all the more dependable in a relatively stationary land-based laboratory. Cytometers that are developed to handle both very small and very large particles may allow flow analysis to move more decisively into the fields of microbiology, parasitology, mycology, and botany. The general idea of studying autofluorescence instead of trying to avoid or ignore it is one that may be profitably considered by cytometrists in many areas of endeavor.

REPORTER MOLECULES

The Herzenbergs' group at Stanford has developed a flow cytometric method for assaying the presence of the enzyme β-galactosidase (coded by the *lacZ* gene from *Escherichia coli*). The presence of this enzyme can be detected by use of a so-called fluorogenic substrate— in this case fluorescein digalactopyranoside (FDG), which is cleaved by β-galactosidase to fluorescein monogalactoside and then to fluorescein, which is fluorescent. The importance of assaying for the presence of β-galactosidase transcends any interest in regulation of expression of this enzyme in bacterial cells: The *lacZ* gene has been used extensively in molecular biology as a reporter for the presence and/or expression of recombinant genes in eukaryotic cells. Cloned genes can be inserted, along with the *lacZ* bacterial gene, into eukaryotic cells; if they are all under the control of the same promoter, expression of the *lacZ* gene will then become a marker for the expression of the cloned genes.

A creative use of this technique was developed by Nolan and Krasnow at Stanford in a system they called *whole animal cell sorting* (WACS). The system does not involve sorting of intact animals (sheep to the left, goats to the right) but rather sorting of all the cells from a whole animal, after they have been dissociated. Specifically, the system has been used to study development of the fruit fly *Drosophila*. Identifiable cell types in developing *Drosophila* embryos have specific promoter regions in their genome that become activated in the course of development to initiate the formation of gene products typical of each cell type. Embryos can be transfected with *lacZ* into chromosome positions driven by a cell-type-specific promoter. These embryos (containing the introduced *lacZ* gene under the control of a specific promoter) are then grown to a given developmental stage. The cells expressing the reporter gene will contain β-galactosidase. Depending on the promoter gene governing the *lacZ* gene, different types of cells will therefore fluoresce when loaded with FDG. The distribution of these fluorescent cells (and therefore the activity of the specific promoter) can be visualized by looking at the intact embryo (Fig. 11.8). The embryos can also, however, be dissociated and the cells expressing the reporter gene then sorted by flow cytometry, based on fluorescein fluorescence intensity (Fig. 11.9). In this way,

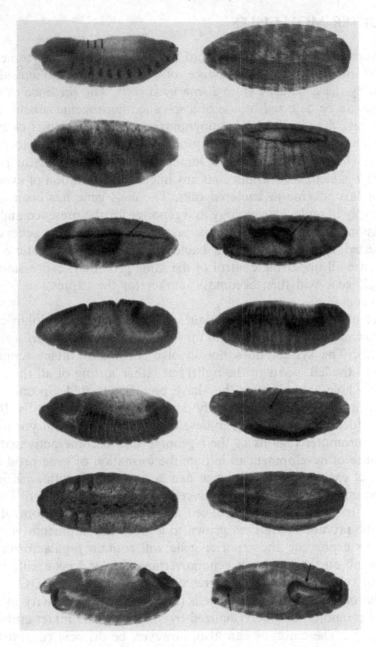

Fig. 11.8. *Drosophila* embryos transfected with the *lacZ* gene into association with different cell-type-specific promoters. Depending on the promoter gene, cells in different patterns over the embryo surface will possess the enzyme β-galactosidase (indicated by dark grains in this photograph). Courtesy of YN Jan from Bier et al. (1989).

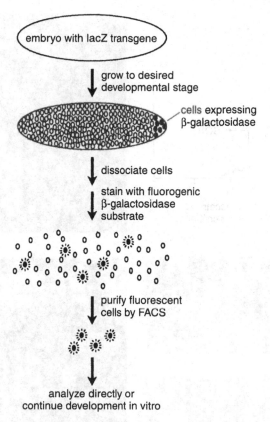

Fig. 11.9. The experimental protocol for whole animal cell sorting (WACS). Courtesy of Mark Krasnow.

cells destined for different functions can be purified and their subsequent development and interactions with other cells observed in culture (Fig. 11.10).

Over the past several years, the jellyfish *Aequorea victoria* has had remarkable impact on the field of reporter molecules in biology. When calcium ions bind to one of its proteins (aequorin), light is emitted. In vitro, aequorin emits blue light when binding calcium. The jellyfish, however, produces green light because a second protein (the imaginatively named "green fluorescent protein" or GFP) receives energy from aequorin and then emits fluorescence at a longer wavelength (remember those tandem fluorochromes). In a landmark paper by Chalfie et al. (1994), GFP was shown to remain fluorescent when transfected into the bacterium *Escherichia coli* (under the con-

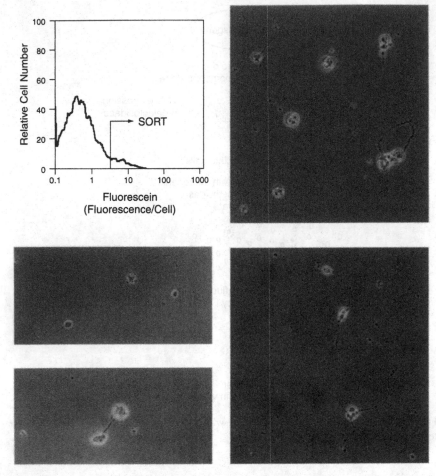

Fig. 11.10. When *lacZ* has been transfected into *Drosophila* embryos in association with a neuronal-cell-specific promoter, the cells that fluoresce brightly in the presence of fluorogenic β-galactosidase substrate will, when sorted, develop neuronal processes in culture. From Krasnow et al. (1991). Science 251:81–85. Copyright AAAS.

trol of an exogenous inducer) and into the nematode *Caenorhabditis elegans* (under the control of a cell-lineage-related promoter). Subsequent work in this field has elaborated a series of GFP-derived fluorochromes of different colors (BFP, CFP, and YFP). They have been used as reporters to detect induction of different proteins or, alternatively, have been engineered as fusion proteins so that, without staining, the presence of these proteins can be assayed by flow cytometry and their cellular location visualized by microscopy. Flow sorting has had an important role to play in the development of stable GFP-fusion protein transfected cell lines (Fig. 9.4 shows an example of sorting for GFP fluorescence).

MICROBIOLOGY

Although microorganisms would seem to be ideal candidates for flow analysis, flow cytometry has been slow to make its presence felt in the field of microbiology. To a great extent, this is attributable to limitations of the instrumentation; flow chambers and sheath fluid and electronics designed for eukaryotic cells have often not worked dependably well with the smaller members of our universe. Without the help of autofluorescent pigments that aid oceanographers in the identification of phytoplankton, microbiologists have difficulty in resolving low-intensity scatter and fluorescence signals from debris and instrument noise. A considerable amount of work has aimed at overcoming some of these instrumental difficulties; many cytometers, if well-aligned, perform acceptably with small bacterial cells.

Much of the work on microorganisms in flow systems has concerned yeast, algae, and protozoa; although smaller than mammalian cells, these eukaryotes are considerably larger than most bacteria. DNA, RNA, protein, and light scatter measurements have been made on these organisms, and the feasibility of cell cycle analysis has been demonstrated. Bacteria, however, present more acute difficulties. The diameter of bacterial cells is perhaps 1 μm (compared with 10 μm for mammalian blood cells), and therefore the surface area to be stained (and resulting fluorescence intensity) is 10^2 less than that of a mammalian cell. The DNA content of the *E. coli* genome is about 10^{-3} times that of a diploid human cell. Hence, bright dyes and sensitive instrumentation are required for studies of bacteria. Nevertheless, reasonable DNA histograms of bacteria can be obtained by flow cytometry. Methods are being developed to investigate cell cycle kinetics, the effects of antibiotics, and the detection and identification of bacteria for clinical investigations.

However, a technique developed at the Massachusetts Institute of

Technology has approached the problem with the "if you can't beat them, join them" philosophy of allowing bacteria to masquerade as larger particles. The technique involves the creation of "salad oil" emulsions of drops of agar within bacterial suspensions in buffer solution. By adjusting the size of the drops and the concentration of the bacteria, it is possible to arrange conditions so that, on average, each drop of gel contains one bacterial cell. The microdrop then becomes a minicontainer for the bacterial cell, allowing the diffusion of stain, nutrient, antibiotics, and so forth, but containing the bacterial cell and all its progeny.

The usefulness of this technique has been shown by its ability to detect the division of these bacterial cells. By staining the cells within the droplets in some way (e.g., for DNA or protein content), the original culture will form a single flow histogram peak representing gel microdroplets, each fluorescing with an intensity related to the DNA or protein content of its entrapped single bacterial cell. After one replication cycle in which all the bacteria are replicating, each droplet will then contain two cells and have twice the original fluorescence intensity. Alternatively, if only some of the bacteria are replicating, a small population of gel droplets with twice the fluorescence intensity will appear. The droplets containing replicating cells will then progress to 4-fold, 8-fold, and 16-fold intensity as replication continues (Fig. 11.11). The technique can provide a sensitive method for studying small particles as well as a very rapid assay for the replication of a small proportion of bacterial cells in the presence of antibiotics, growth factors, or varied growth conditions (Fig. 11.12).

Although this gel microdroplet method is still new (even after 10 years) and its potential applications relatively untested, it has been described here because it can teach us certain general lessons. It serves to remind us that cytometry, despite its name, does not necessarily involve the flow analysis of cells; particles of many sorts will do just as well. It is also of interest as a method that has, in fact, institutionalized the formation of clumped cells that most workers try so hard to avoid. In addition, it has provided us with a way to make a small cell into a larger (flow-friendly) particle. Finally, it has given us inspiration by exemplifying the way in which lateral thinking can extend the impact of flow cytometry in new directions.

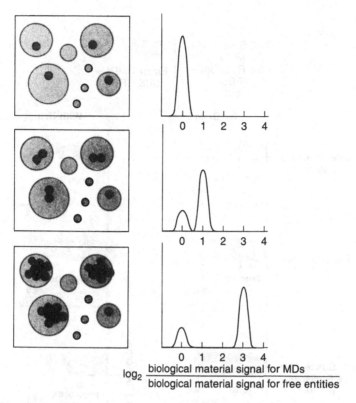

$$\log_2 \frac{\text{biological material signal for MDs}}{\text{biological material signal for free entities}}$$

Fig. 11.11. Illustration of the use of gel microdroplets for sensing growth at the level of one cell growing into a two-cell microcolony. By staining the cells within microdroplets with, for example, a DNA-specific fluorochrome, a small subpopulation of cells dividing more or less rapidly than most could be detected in a flow histogram of microdroplet fluorescence. From Weaver (1990).

MOLECULAR BIOLOGY

Flow cytometry has been applied in many creative ways to the science of molecular biology. Flow sorting, based on Hoechst 33258 and chromomycin A3 fluorescence, turns out to be one of the best ways available for obtaining relatively pure preparations of each type of chromosome. Even those chromosomes that are not distinguishable by their fluorescence (e.g., 9-12) can usually be sorted from hamster–human hybrid cell lines. These preparations of reasonably

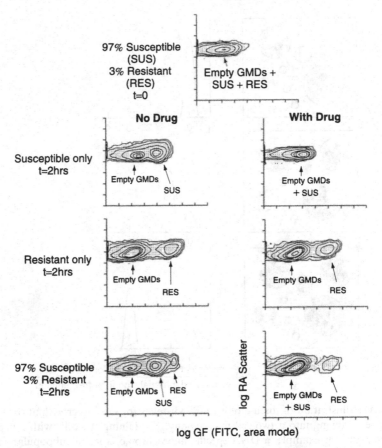

Fig. 11.12. The use of gel microdroplets and flow cytometry to assay drug sensitivity of bacterial cells. The figure shows side scatter and green fluorescence contour plots of gel microdroplets (GMDs) containing *E. coli* cells that have been stained with fluorescein isothiocyanate for total protein. The microdroplets have been analyzed in the flow cytometer either at time 0 or 2 h after incubation in control medium (left plots) or medium containing penicillin (right plots). A model system was created by mixing two strains of bacteria (susceptible or resistant to penicillin). The data show that a small subpopulation of resistant cells could be detected within 2 h because of its rapid growth in comparison to susceptible cells. From Weaver et al. (1991).

pure flow-sorted chromosomes have been the starting material for obtaining chromosome-specific DNA libraries. The Human Genome Project has made extensive use of chromosomes obtained with the sorting cytometers at Los Alamos and Livermore. The bottleneck in

TABLE 11.1. Chromosome Sorting

Application	Chromosomes required
Polymerase chain reaction chromosome paints	3.0×10^2
Viral cloning (17 kb inserts)	1.6×10^6
Cosmid cloning (37 kb inserts)	6.4×10^6
Yeast artificial chromosome cloning	6.4×10^7

Sorter	Chromosomes sorted per 24 h day	Estimated sorting time for YAC cloning (6.4×10^7 chromosomes)
Conventional	3×10^6	21 days
High speed	1.5×10^7	4 days
Optical zapper	7.5×10^7	0.8 days

After Roslaniec MC et al. (1997). Hum. Cell 10: 3–10.

this technique is the time it takes for sorting. In a conventional sorter, if the flow rate is limited to about 1000 particles per second, then human chromosomes of a particular type could theoretically be sorted at a rate of about 40 per second (1000/23). Thus it would take about 7 h to obtain 10^6 chromosomes of a given type under optimal conditions. The high-speed drop sorters at Los Alamos and Livermore were developed with the demands of the Human Genome Project in mind. The technique for optical (zapper inactivation) sorting is the next step in this evolution (Table 11.1).

Even small numbers (20,000 or so) of sorted chromosomes can be used quite neatly to aid in the mapping of genes to chromosomes. Chromosomes of each type are simply sorted (two at a time: one type to the left, the other to the right) onto a nitrocellulose filter. The DNA is then denatured on the filter where it can be hybridized to a radioactive gene probe. Autoradiography of the filter will then reveal whether the probe has hybridized to the DNA from any given chromosome (Fig. 11.13). In this way, the sorting of small numbers of each of the chromosomes onto filters allows the mapping of any available gene probe to its chromosome.

Much like chromosomes, DNA fragments generated by restriction enzyme digestion of native DNA, when stained with DNA-specific

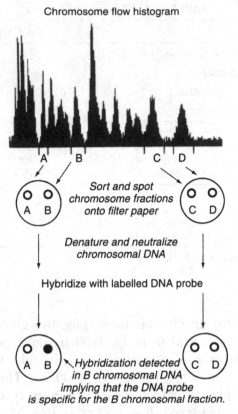

Fig. 11.13. DNA blot analysis using spots from chromosomes sorted directly onto filter paper on the basis of their Hoechst 33258 fluorescence. From Van Dilla et al. (1990).

fluorochromes, fluoresce more or less brightly according to their length. Flow data acquisition could, in theory, provide histograms indicating the relative proportions of different-sized fragments in a digest. Implementation of this application has not been trivial because, compared with chromosomes, fragments of DNA are much smaller and fluoresce much less brightly. Rising to this challenge, scientists have developed ultrasensitive flow cytometric systems by using high-efficiency optics and light detection, bright fluorochromes, and very slow sheath velocities of approximately 2–4 cm per second (compare this with the usual cytometers with velocities of 10 m [1000 cm] per second). With slow flow rates, DNA fragments spend longer

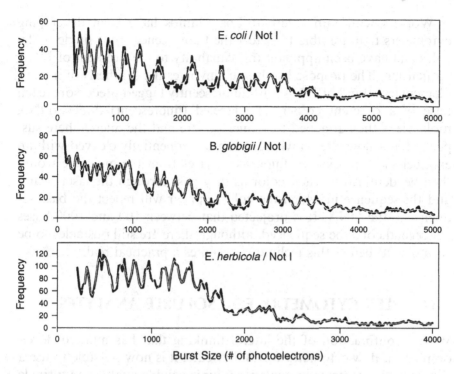

Fig. 11.14. Flow cytometric signatures distinguishing three different strains of bacteria according to the size distributions of DNA fragments generated by restriction enzyme digestion. From Kim et al. (1999).

time in the laser beam, thereby emitting more fluorescence per fragment and increasing instrument sensitivity for detection of short fragments with low intensities. As an alternative to pulsed-field gel electrophoresis, the flow methodology records similar information about the distribution of fragment sizes, but with better resolution, greater accuracy, and higher sensitivity. In addition the flow method requires less DNA and less time. In one proof-of-principle application of this technology, when DNA from bacteria has been broken into fragments by digestion with restriction enzymes, the size distribution of the fragments is typical of a given strain of bacteria. Therefore the flow histogram that is derived from sending the fragments through a flow cytometer is a signature of the particular bacterial strain and can be used for identification (Fig. 11.14).

Workers with James Jett at Los Alamos have been developing cytometers that are able to detect the fluorescence from single molecules and have been applying this capability to the problem of DNA sequencing. The proposed technique involves the synthesis of a complementary strand of DNA with fluorescently tagged precursors (each base of a different color). The labeled (fluorescent) duplex DNA molecule is then attached to a microsphere and the microsphere suspended in a flow stream where it can be sequentially cleaved with an exonuclease. The cleaved fluorescent bases from the molecule would then be identified by their color as they pass through the laser beam, and the sequence in which the colors appear will reflect the base sequence in the DNA. It is projected that between 100 and 1000 bases per second could be sequenced, although there are still obstacles to be surmounted before this technique becomes a practical reality.

MULTIPLEX CYTOMETRY FOR SOLUBLE ANALYTES

With a continuation of the lateral thinking that has marked developments in flow cytometry for many years, it is now possible to use a simple flow cytometer to analyze multiple soluble analytes in a single test tube. Sets of beads can be manufactured with distinctive ratios of two different fluorochromes (for example, red and orange). By precisely controlling the ratios of these two dyes, bead sets can be obtained with 100 different specifications (according to their red/ orange balance). When run on a flow cytometer, the beads cluster into 100 separate regions on a two-color (red vs. orange) dot plot (Fig. 11.15). In a multiplexed version of an ELISA assay, each bead type can be linked to a different capture molecule; the capture molecule will bind a soluble target analyte to the bead (compare this with the solid-phase capture molecules on an ELISA plate). For example, capture molecules on the beads might be a range of allergens (to bind antibodies from the serum for allergy profiling) or nucleotide probes for single-nucleotide polymorphisms (SNPs) or antibodies to a range of cytokines.

After preparation (with beads of each red/orange ratio having been linked to capture molecules for specific targets), the beads can all be combined in a single tube. For the assay, a test solution is added to the mixed bead suspension. After incubation, centrifugation,

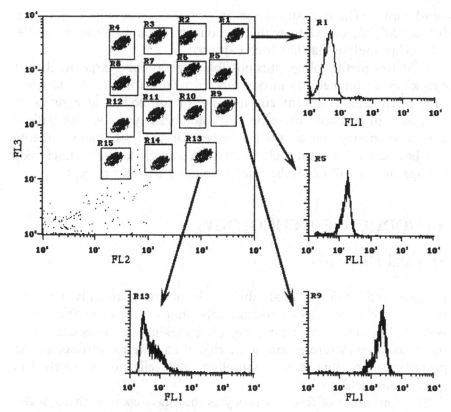

Fig. 11.15. A multiplex bead array that can be used to capture multiple soluble analytes. In the dot plot, 15 types of beads are distinguished by their red/orange ratios; they capture 15 different cytokines. For example, the beads in Region 1 capture interleukin (IL)-9; in Region 5, IL-2; in Region 9, IL-5; and in Region 13, MCP-1. The green fluorescence histograms from each region show, by their intensity, how much cytokine has been captured by each type of bead. Standard curves can relate the green fluorescence intensity to the concentration of cytokine in the sample. The experimental format illustrated in the cartoon here is patterned directly from work by RT Carson and DAA Vignali (1999).

and washing, fluorescein-conjugated detection reagents are added to the incubation mixture. Much like the top layer in an ELISA assay, the detection reagents will bind to the target analytes on the beads. After running the tube of beads through the flow cytometer, the final read out comes from gating on each one of the 100 types of beads in turn and then (knowing the particular capture antibody on that type of bead) reading the green fluorescence intensity on that red/orange bead cluster. The intensity of the green fluorescence will, under careful conditions, correlate with the concentration in solution of the particular analyte that has been captured.

With this methodology, it is possible to assay for a large number of cytokines simultaneously in one tube or to test for a complete selection of clinically relevant enzymes or for antibodies to a series of allergens. In addition, by using one color bead type for each patient and then mixing the samples together for analysis, many patients can be assayed for a particular analyte in a single test tube. Each red/orange cluster will mark the results from a particular patient.

REPRODUCTIVE TECHNOLOGY

Prenatal Diagnosis

Prenatal diagnosis of fetal abnormalities conventionally involves testing of cells from either the amniotic fluid or the chorionic villus. Both techniques involve sampling of tissues closely associated with the fetus; they therefore pose some risk of causing miscarriage of the pregnancy. For this reason, sampling fetal cells noninvasively has been a long-held goal.

Small numbers of fetal erythrocytes that have escaped through the placenta exist in the maternal circulation. These cells can be detected by their fetal hemoglobin (which distinguishes them from most maternal erythrocytes). Their presence in significant numbers (greater than 0.6%) has been used to diagnose a fetal–maternal hemorrhage across the placenta and to mandate clinical intervention to prevent the development in the Rh-negative mother of anti-Rh antibodies (see Chapter 10). Because fetal cells are accessible by drawing blood from the mother, sampling of them from the mother's circulation involves no hazard to the fetus and holds the possibility of noninvasive prenatal diagnosis of fetal abnormalities. However, prenatal diagnosis requires DNA, and most of the fetal erythrocytes in the maternal circulation are not nucleated. Over the past decade, work in this field (led, to a great extent, by Diana Bianchi at Tufts University in Boston, Massachusetts) has centered on the development of methods for distinguishing nucleated fetal cells from maternal cells and from non-nucleated fetal cells and then on methods for isolating these cells for subsequent fluorescence in situ hybridization (FISH) analysis of the fetal genome for disease-related sequences. Because the nucleated fetal cells are very rare (estimated to be fewer than 20 nucleated fetal erythrocytes per 20 ml of maternal peripheral blood; that is, a cell ratio of about $1:10^9$), flow sorting has been proposed as a method for purifying or at least enriching the population before FISH analysis.

Such sorting requires a flow-detectable method for classifying nucleated fetal cells. Two protein markers, among many others, that have been proposed to distinguish fetal cells from maternal cells are the transferrin receptor and fetal hemoglobin. Antibodies against fetal hemoglobin have been used (by Bianchi) in conjunction with a Hoechst stain for DNA to detect only those cells that have fetal hemoglobin and are also nucleated. As a model system for the reliability of flow sorting to enrich for fetal cells, blood from women carrying male fetuses has been studied. The presumptive fetal cells from the maternal blood are sorted (based on Hoechst and anti-fetal hemoglobin staining) and then tested with chromosome-specific FISH probes for the presence of a Y chromosome or for only a single X-chromosome (proving that the sorted cells were, indeed, from the male fetus and not from the mother). Results from these experiments are promising. Although these techniques are still at early stages of investigation, there is considerable optimism that flow sorting of fetal cells may fulfill the need for prenatal diagnosis of fetal abnormalities without the ethical problems posed by invasive sampling of low risk pregnancies.

Determining Sex

In another facet of reproductive technology, flow sorting has been used to separate sperm bearing X and Y chromosomes. If you own a dairy herd, you want your cows to produce more cows (and few bulls). Conversely, beef farmers want males. Because bovine X chromosomes are considerably larger than Y chromosomes, sperm bearing X chromosomes have approximately 4% more DNA that their Y bearing brothers. By staining with a Hoechst dye, the two classes of sperm can be sorted, with a flow cytometer, according to their fluorescence intensity. Sex determination of the offspring, resulting from insemination with the sorted sperm, is not perfect (think of the possibility of overlap between the intensities of sperm that are separated by only 4%), but it is not bad. There is now a commercial flow cytometer that has been developed specifically for sorting animal sperm. Sex-selected calves are being produced more or less routinely by artificial insemination with sorted sperm. The technology has also been applied to the high-stakes world of race horse breeding.

Speaking of high-stakes worlds, there is now at least one commercial clinic that provides sperm sorting for humans ("for prevention of genetic disorders" or "to increase their probability of having a child of the less represented sex in the family"). Although human sperm bearing X and Y chromosomes differ in DNA content (and fluorescence intensity) by only 2.8%, the technique does appear to work with some reliability. In a 1998 summary from the clinic, 14 pregnancies from insemination with X-bearing sperm were reported: 92.9% of those were female as desired. The procedure is not inexpensive, it may have arguable ethical implications, and it is not perfect. The question remains as to whether it could be a glimpse of the future. If it is, then we need to ponder the fact that most American families seem to want daughters....

FURTHER READING

Section 9 in **Current Protocols in Cytometry** provides methods and some theory on many flow cytometric functional assays (including those mentioned here and many others). Chapter 20 in **Darzynkiewicz** is a discussion of methods related to staining lymphocytes for activation antigens.

A special issue of Cytometry (Vol. 10, No. 5, 1989) is devoted to "Cytometry in Aquatic Sciences." A more recent issue of Scientia Marina (Vol. 64, No. 2, 2000) is also devoted to the same subject: "Aquatic Flow Cytometry: Achievements and Prospects." In addition, Chapter 31 of **Melamed et al.** discusses the application of flow cytometry to higher plant systems.

Section 4 in **Diamond and DeMaggio** has several methodological chapters discussing reporter genes, GFP, and cell tracking dyes. The classic reference on GFP is by Chalfie M, et al. (1994). Green fluorescent protein as a marker for gene expression. Science 263:802–805. A good review of the uses of flow cytometry for GFP analysis is by Galbraith DW, Anderson MT, Herzenberg LA (1999). Flow cytometric analysis and FACS sorting of cells based on GFP accumulation. Methods Cell Biol. 58:315–339.

Discussion of the WACS methodology can be found in Krasnow MA, Cumberledge S, Manning G, et al. (1991). Whole animal cell sorting of *Drosophila* embryos. Science 251:81–85.

The gel microdroplet technique is described in articles by Powell KT, Weaver JC (1990). Gel microdroplets and flow cytometry: Rapid determination of antibody secretion by individual cells within a cell population. Bio/Technology 8:333–337; and Weaver JC, Bliss JG, Powell KT, et al. (1991). Rapid clonal growth measurements at the single cell level. Bio/ Technology 9:873.

Research on flow cytometry of microorganisms is reviewed in Chapter 29 of **Melamed et al.** and in Section XI in **Darzynkiewicz.** Microbiological methods are described in Section 11 of **Current Protocols in Cytometry.** A useful book edited by D Lloyd (1993) is Flow Cytometry in Microbiology (Springer-Verlag, London).

A good reference on the flow cytometry of DNA fragments is Kim Y, Jett JH, Larson EJ, et al. (1999). Bacterial fingerprinting by flow cytometry: Bacterial species discrimination. Cytometry 36:324–332. The proposed technique for sequencing DNA is described in a paper by Jett JH, Keller RA, Martin JC, et al. (1989). High-speed DNA sequencing: An approach based upon fluorescence detection of single molecules. J. Biomol. Struct. Dynamics 7:301–309.

Results using multiplexed bead flow cytometry have been reported by Carson RT, Vignali DAA (1999). Simultaneous quantitation of 15 cytokines using a multiplexed flow cytometric assay. J. Immunol. Methods 227:41–52.

A review of progress in developing flow methods for prenatal diagnosis is by Bianchi DW (1999). Fetal cells in the maternal circulation: Feasibility for prenatal diagnosis (review). Br. J. Haematol. 105:574–583. In addition, Chapter 20 by JF Leary in **Darzynkiewicz, 1994** discusses general issues of rare cell cytometry as well as specific reference to the sorting of fetal cells.

A provocative, nontechnical article on the ability to choose, with the aid of a flow cytometer, the sex of our children was written by L Belkin ("Getting the Girl" in The New York Times Magazine, July 25, 1999, pp 26–55). Results from a clinic that sells this technique to prospective parents have been reported by Fugger EF, Black SH, Keyvanfar K, Schulman JD (1998). Births of normal daughters after MicroSort sperm separation and intrauterine insemination, in-vitro fertilization, or intracytoplasmic sperm injection. Hum. Reprod. 13:2367–2370.

12

Flowing On: The Future

In the Preface, I referred to the way flow cytometry has moved over the past three decades in directions that have surprised even those intimately involved in the field. It is obvious that attempts to predict the future of flow cytometry should be left to informal discussions among friends. The temptation to speculate is, however, an impossible one to resist. This chapter should therefore be read in the spirit in which it was written: with a large measure of skepticism (preferably among friends, after a good dinner and a good bottle of wine).

The capabilities of state-of-the-art research cytometers should, over the next years, continue to move toward easier analysis of particles of greater size range with more sensitivity and greater speed than they do now. Although currently including particles from over 100 μm (like plant protoplasts) down to bacteria and DNA fragments on research instruments, this range may soon become possible on benchtop cytometers. Exploitation of this flexibility may finally lead to greater use of flow cytometers for looking at plant cells and also at microbiological systems. It may also be possible to use cytometers to look at the properties of isolated organelles such as mitochondria and chloroplasts. It is likely that some type of flow cytometry will begin to be used more routinely in bacteriology and environmental laboratories. Recent developments with multiplexed bead-based assays lead many to guess that flow cytometry is also poised to enter the biochemistry laboratory, as an alternative to the ELISA technique for assaying soluble components.

High-speed drop sorters may continue to increase in speed, although the high pressures required in these sorters may have reached their limit in terms of the viability of the sorted cells. Optical sorting (by using a laser to inactivate unwanted cells or chromosomes) may begin to provide a practical, rapid alternative to drop sorting. An obvious method for allowing more effective use of time and sample on conventional drop sorters would be to sort four or more populations (instead of just two) simultaneously. Although slow to catch on, four-population sorting is now available on commercial instruments. This involves charging some drops more than others so that drops can be deflected slightly to the left and to the right as well as strongly to the left and to the right—giving four options and allowing four populations to be sorted during each run. Four populations is not the technical limit—if biologists can think of experiments that demand more.

Extra lasers have been appearing even now on benchtop cytometers; excitation lines may continue to become less limiting in the choice of fluorochromes and the design of experiments. Developments in fluorochrome technology, in conjunction with extra lasers, have already facilitated super-multiparameter analysis by allowing the use of wider regions of the spectrum without unsurmountable problems of spectral overlap. Whereas routine instruments today look at forward and side scatter as well as three or four fluorescence parameters, some high-tech instruments already quite happily look at 11 fluorescence parameters (plus forward and side scatter), and, with demand, the electronics are capable of handling even more. The question really concerns, then, the ability of the human intellect to plan rational experiments and to cope with the information that can be obtained in a short period of time from particles that are analyzed for large numbers of parameters. Even after more than a decade of multiparameter benchtop cytometry, it is not clear that our ability to design meaningful experiments with multiple measured parameters has yet caught up with our ability to measure those parameters. Despite the ability to look simultaneously at three or four fluorescent stains quite routinely on many instruments, most people still seem quite happy to stick to two. Whether this is related to general conservatism, the high cost of additional antibodies, problems with cross-reactivity in complex staining systems, or some inherent human discomfort with forays into multidimensional reasoning is a question that may be answered in the future.

There may well be increasing interest in analysis of the intrinsic characteristics of cells that lead to alterations in autofluorescence. John Steinkamp, Harry Crissman, and others at the Los Alamos Laboratory have been using flow systems to study the time characteristics (that is, the decay kinetics) of fluorescence emission, as a handle for studying alterations in the microenvironment of fluorochromes bound to molecules in the cell; their results are exciting and might be used to study autofluorescence in the future. On a larger time scale, biochemical kinetics experiments may become more popular with increasing use of functional probes. It may also happen that flow cytometrists and microscopists will begin, at last, to talk to each other when they discover that they share common interests in image analysis microscopy. Laser-scanning cytometry already marks the beginning of methodological communication between the analysis strategies typical of flow cytometry and the microscope-based comfort zone required by pathologists. A cytometry laboratory of the future may well be equipped with a collection of fluorescence and confocal microscopes, laser-scanning cytometers, and flow cytometers—all being used, as appropriate, by the same people.

The extra information collected by rapid cytometers looking at many parameters will certainly increase the demand on data storage systems. Ten years ago, when writing the first edition of this book, I said that, against general scientific practice, flow cytometrists might need to begin to learn to wipe out data that they no longer need. Less expensive media for data storage have now made it possible to avoid such measures. Zip cartridges and CD-ROMs are the media of choice now; DVDs may be the choice in the near future (and the next "perfect" storage medium may already be on someone's drawing board). Given the easier availability of inexpensive storage media, I see no reason to expect that our demands for data storage capacity will do anything other than continue to increase. With regard to software, the trend may continue toward a healthy proliferation of varied packages for flow analysis with full compatibility of data acquired on any and all systems.

I said at the beginning of this book that flow cytometry is currently moving in two directions at once: Technological advances provide, in one direction, increasingly rapid, sensitive, complex, but precarious analysis and, in the other direction, increasingly stable, fool-proof, and automated capabilities. Thirty years ago, all flow cytometry was at the complex, but precarious level of development. Now, many of those early precarious developments have been incorporated into routine cytometers, and new techniques are appearing in research instruments. Over the next few years, this progression will continue:

The vast majority of future flow cytometric analysis will be done using routine, black-box instruments of increasingly greater capabilities. In association with this trend, expert (artificial intelligence) systems may begin to have an impact. Robotic systems, incorporating staining, lysing, and centrifugation steps, may allow staff to put blood or other crude material in at one end and get formatted print-outs of flow data from the other. A remaining question is whether fluidics stability, reagent quality control, and subjectivity in gating will ever allow us to relax with this level of automation.

And what does the future hold for the *spirit* of flow cytometry? Although there will continue to be many new technical developments (some predicted and some surprising), I suspect that the sense of adventure may be lost from the field as the balance continues to shift from innovative technology to routine use. The trade-off for that loss of excitement could be the satisfaction that people who understand flow cytometry may feel from having played a part in the comfortable acceptance of this powerful technique by the broad scientific community.

FURTHER READING

Cualing HD (2000). Automated analysis in flow cytometry. Cytometry 42:110–113.

Durack G, Robinson JP, eds. (2000). Emerging Tools for Single Cell Analysis: Advances in Optical Measurement Technologies. Wiley-Liss, Inc, New York.

Roslaniec MC, Bell-Prince CS, Crissman HA, et al. (1997). New flow cytometric technologies for the 21st century. Hum. Cell 10:3–10.

General References

Although books on flow cytometry abound and articles on flow cytometry can be found throughout a great range of publications, the following is a limited list of references that I have found particularly useful for general information on the theoretical basis of flow analysis and as routes into the literature on particular subjects and techniques.

BOOKS

Bauer KD, Duque RE, Shankey TV, eds. (1993). Clinical Flow Cytometry: Principles and Application. Williams & Wilkins, Baltimore. A multiauthor compendium of flow cytometry in clinical practice.

Darzynkiewicz Z, Robinson JP, Crissman HA, eds. (2001). Methods in Cell Biology, Vols 63A and 63B. Cytometry, 3rd edition. Academic Press, San Diego. An up-to-date compilation covering theory and many practical aspects of flow and image cytometry. The previous edition of this book (Vols. 41A and 41B, 1994) contains many articles on many subjects not covered in the third edition and is also worth reading.

Diamond RA, DeMaggio S, eds. (2000). In Living Color: Protocols in Flow Cytometry and Cell Sorting. Springer, Berlin. A collection of detailed methods, covering all aspects of flow cytometric analysis.

Keren DF, Hanson CA, Hurtubise PE, eds. (1994). Flow Cytometry and Clinical Diagnosis. American Society of Clinical Pathologists Press, Chicago. A good summary of issues related to the practice of clinical cytometry.

Melamed MR, Lindmo T, Mendelsohn ML, eds. (1990). Flow Cytometry and Sorting, 2nd edition. Wiley-Liss, Inc, New York. A classic and very thick (824 page) multiauthor volume, containing thorough review articles on the theory and practice of flow cytometry by acknowledged experts.

Ormerod MG, ed. (2000). Flow Cytometry: A Practical Approach. Oxford University Press, Oxford. A useful book with discussion of flow theory but with an emphasis on practical aspects of flow cytometry and technical protocols.

Ormerod MG (1999). Flow Cytometry. Bios Publishers, Oxford. Short and sweet.

Owens MA, Loken MR (1995). Flow Cytometry: Principles for Clinical Laboratory Practice. Wiley-Liss/John Wiley and Sons, Inc, New York. A discussion of the theory and practice of clinical flow cytometry, with some good emphasis on quality control.

Robinson JP, ed. (1993). Handbook of Flow Cytometry Methods. Wiley-Liss/John Wiley and Sons, Inc, New York. An informal, detailed manual of techniques, protocols, and supplies that are relevant to flow cytometric analysis.

Robinson JP et al., eds. (2000). Current Protocols in Cytometry. John Wiley and Sons, Inc, New York. An ever-increasing volume (in ring-binder/loose-leaf or CD format) with articles on the theory of flow cytometry and image analysis as well as excellent discussions on the practical methodologies. It is kept current with quarterly supplements.

Shapiro HM (1995). Practical Flow Cytometry, 3rd edition. Wiley-Liss/John Wiley and Sons, Inc, New York. This wonderful book manages to make learning about flow cytometry more enjoyable than you would have thought possible. There are readable sections on most aspects of flow theory and an excellent compendium of references. A fourth edition is, I believe, in the works.

Stewart CC, Nicholson JKA, eds. (2000). Immunophenotyping. Wiley-Liss/John Wiley and Sons, Inc, New York. A multiauthor volume with an emphasis on clinical leukemia and lymphoma phenotyping, but with useful articles on general phenotyping issues, on

HIV infection, CD34 stem cell enumeration, platelets, and transplantation cross-matching.

Van Dilla MA, Dean PN, Laerum OD, Melamed MR, eds. (1985). Flow Cytometry: Instrumentation and Data Analysis. Academic Press, London. A venerable, but still current book with an emphasis on the physics and mathematics of flow systems and data analysis. It has some excellent (and readable) articles on some theoretical subjects.

Watson JV (1991). Introduction to Flow Cytometry. Cambridge University Press, Cambridge. A somewhat idiosyncratic tour through the many theoretical aspects of flow cytometry with which Watson is well-acquainted. There are detailed discussions of the limits on signal resolution, on cell coincidence in the laser beam, and on methods for looking at dynamic cell events. There is also good coverage of oncological applications but no mention at all of lymphocytes.

Watson JV (1992). Flow Cytometry Data Analysis: Basic Concepts and Statistics. Cambridge University Press, Cambridge. A detailed look at the ways we can (or should) analyze data—after we leave the flow cytometer.

Weir DM, ed. (1986). Handbook of Experimental Immunology, Vol 1: Immunochemistry. Blackwell Scientific Publications, Oxford. A detailed reference volume on many aspects of immunology, including immunofluorescence techniques and antibody conjugation methods as well as flow cytometric analysis. A newer edition (1996) of Weir's Handbook of Experimental Immunology (edited by Leonore Herzenberg) is also available—in four volumes.

CATALOGUES

The following are catalogues and handbooks from manufacturers. They are free and, even at ten times that price, would be well worth owning. Most are available in either paper or CD-ROM versions.

Melles Griot Catalog (http://www.mellesgriot.com/). This paper or CD guide provides a great deal of theoretical information about filter, lens, mirror, and laser specifications and design as well as about the Melles Griot range of products.

Molecular Probes: Handbook of Fluorescent Probes and Research Chemicals, by Richard P. Haugland. Molecular Probes (http://www.molecularprobes.com/) makes a vast range of fluorescent chemicals that are useful in flow cytometric analysis. The website and the paper or CD Handbook provide a great deal of information about the use of these chemicals as well as about their photochemical characteristics. A required reference book for every flow cytometrist.

Hamamatsu Corporation (http://www.usa.hamamatsu.com/). This website contains a good discussion of photomultiplier tube electronics. We all need to be reminded occasionally that a flow cytometer's performance is never any better than the performance of its photodetectors.

Boehringer Mannheim Biochemicals publishes a very useful manual on "Apoptosis and Cell Proliferation."

MISCELLANEOUS

The journal that specializes in research reports about flow techniques and flow analysis (as well as image analyis) is *Cytometry*. Its paired publication is *Communications in Clinical Cytometry*. The *Journal of Immunological Methods* is also often useful in this regard.

The International Society for Analytical Cytometry (ISAC, 60 Revere Drive, Suite 500, Northbrook, IL 60062; http://www.isac-net.org) is the society that specializes in flow and image cell analysis. Meetings (attended by about 1000 people) are held every 2 years (sometimes in the United States and sometimes in Europe).

The U.S. National Flow Cytometry Resource is at the Los Alamos National Laboratory (Los Alamos, NM 87545). They are a source of information and inspiration and provide help and facilities for scientists wanting to make use of their "state-of-the art" cytometers. They also run flow cytometry training courses.

Websites on flow cytometry reflect the idiosyncracies of their authors. Many are excellent. For current listings of this ever-changing scene, search the Purdue University Cytometry Laboratories website (http://www.cyto.purdue.edu/) or the ISAC web site (http://www.isac-net.org/).

There is an e-mail network for informal discussion on flow issues. It has, as of this writing, 1999 members. Participants in this network are a singularly helpful and generous group of people. The network is run by Paul Robinson and Steve Kelley from Purdue University. Sign up by writing to Steve Kelley (cyto-request@flowcyt.cyto.purdue.edu).

Courses on flow cytometry are held at various locations, mainly during the summer. For current listings check the Purdue website (http://www.cyto.purdue.edu/), which is usually as up-to-date as possible.

Glossary

There is a fine line between words that provide necessary technical information and words that we might call *jargon*. Whereas technical vocabulary is important as a means to intellectual precision, jargon can often be used either to obscure ignorance or, like a badge, to identify members of an exclusive club. Both technical vocabulary and jargon, however, form a barrier between people already within a field of endeavor and those attempting to enter that field. Without identifying which of the following words are necessary and which are merely jargon, I include this somewhat selective glossary as an effort toward lowering that barrier.

Absorption: In the context of photochemistry, *absorption* refers to the utilization, by an atom or molecule, of light energy to raise electrons from their ground-state orbitals to orbitals at higher energy levels. Having absorbed the light energy, the atom or molecule is now in an excited state and will emit energy (in the form of either heat or light) when it returns to its ground state. Atoms will absorb light if, and only if, it is of a wavelength whose photons contain exactly the amount of energy separating a pair of electron orbitals within that atom.

Acquisition: In flow cytometry, *acquisition* refers to the process of recording the intensity of the photodetector signals from a particle in the transient memory of a computer. Once acquired, the data from a group of particles can be stored permanently on a storage medium from which it can be subjected to analysis. Acquisition and then analysis (in that order) are the two central steps in the flow cytometric procedure.

Acridine orange: Acridine orange (AO) is a stain that fluoresces either red or green, depending on whether it is bound to double-stranded or single-stranded nucleic acid. It has proved useful in comparing DNA and RNA content within cells; and it has also been used successfully (in the presence of RNase and mild denaturing conditions) to look at the changes in DNA denaturability during the cell cycle. There is debate in the flow cytometric community about whether the reputation that AO has for being difficult to work with is justified.

Activation marker: Activation markers are proteins that come and go on the surface of cells in response to stimulation. As such, they provide functional information about the physiological state of the cell. They also provide flow cytometrists with a reason to be concerned about instrument stability, sensitivity, and standardization.

ADC: An analog-to-digital converter converts photodetector signals (called *analog signals* because they are continuously variable, having an infinite variety of values) to channel numbers. The light intensity range represented by a given ADC channel depends on the amplification applied to the photodetector signal. ADCs can have 256 or 1024 or even 65,536 channels.

Aerosol: An aerosol is the spray of small fluid droplets that can be generated particularly when the nozzle of a flow cytometer is vibrated for sorting applications. If samples contain material that may be a biological hazard, attention should be paid to containment of the aerosol by suction through small-pore, hydrophobic filters.

Algae: Algae are simple forms of plant life. The larger algae are known as *seaweeds*. The unicellular algae form a large part of the plankton of both marine and fresh water environments and are suitable for analysis by flow cytometers. Algae also have contributed greatly to flow cytometric analysis because of their elaboration of pigments like phycoerythrin, PerCP, and allophycocyanin, which can be conjugated to antibodies and which facilitate multicolor staining procedures because they fluoresce in different regions of the spectrum.

Analysis: After acquisition, data from a sample are processed so as to provide useful results. This processing, or analysis, consists primarily of the correlation, in some or all possible ways, of the intensity channels from each of the signals recorded for each of the particles acquired for a given sample.

Analysis point: The three-dimensional volume in space where the laser beam intersects and illuminates the sample core of the fluid stream is the analysis point. The size of this point is determined by the cross-sectional dimensions of the laser beam and the width of the stream core itself. It is within the volume of the analysis point that particles are illuminated and signals are detected.

Aneuploid: Although *aneuploid* is used by cytogeneticists to refer to cells with abnormal numbers of chromosomes, it has been hijacked (with intellectual imprecision) by flow cytometrists to refer to the characteristic of possessing an abnormal amount of DNA. To be precise, flow cytometrists should use the term *DNA aneuploid* to acknowledge this distinction.

Annexin V: Annexin V is a molecule that binds to phosphatidylserine and, therefore, if conjugated to a fluorochrome, will identify apoptotic cells (which express phosphatidylserine on their surface). In the assay for apoptosis, annexin V must be used in conjunction with propidium iodide in order to exclude dead cells (which express phosphatidylserine on the internal side of their cytoplasmic membranes).

Apoptosis: Apoptosis is an ordered, active process that brings about the death of a cell as an important part of the maintenance of organismal homeostasis. Apoptosis can be assayed, in flow cytometry, by, for example, looking at the expression of phosphatidylserine on the cell surface, by looking for nuclei with less-than-normal (sub-G0/G1) amounts of DNA, and by looking for an increase in DNA fragment termini.

Arc lamp: An arc lamp is a device that emits light from a gas discharge between two electrodes. The wavelength of the light is determined by the gas used.

Autofluorescence: The light emitted naturally by an unstained, illuminated cell is called *autofluorescence*. The amount of autofluorescence will differ depending on the type of cell, on the wavelength of the illuminating light, and on the fluorescence wavelength being analyzed. Autofluorescence, in general, results from endogenous compounds that exist within cells. It can be studied as an interesting phenomenon in itself; however, bright levels of autofluorescence at a particular wavelength will lower the sensitivity of a flow system for detecting positive stain of low intensity.

Back-gating: Back-gating is a strategy by which the fluorescence of stained cells within a mixed population is used to determine the side scatter and forward scatter characteristics of those cells. This is in contrast to "traditional" gating, whereby scatter characteristics are used to delineate cells whose fluorescence characteristics are then determined.

Beads: Beads are particles (made, usually, of polystyrene) that can be used as stable and inert standards for flow cytometric analysis. Beads can be obtained conjugated to various fluorochromes in order to standardize fluorescence detection settings and optical alignment or to calibrate fluorescence scales. They can also be conjugated to antibodies in order to calibrate the scale in terms of number of binding sites. More recently, in multiplexed assays, beads with capture molecules have been used to determine the concentration of soluble analytes.

BrdU: 5-Bromodeoxyuridine (also abbreviated BUdR or BrdUrd) is a thymidine analog that will be incorporated into the DNA of cycling cells. Cells pulsed with BrdU can then be stained with anti-BrdU monoclonal antibodies to indicate which cells have been synthesizing DNA during the pulse period. BrdU staining is a more precise way to look at the proportion of cells in S phase than simple propidium iodide staining for DNA content.

Break-off point: At the break-off point, some distance from a nozzle orifice, a vibrating stream begins to separate into individual drops that are detached and electrically isolated both from each other and from the main column of the stream.

Calcium: This ion's rapid flux between cellular compartments and from the external medium is important in signal transduction within cells. Its concentration can be measured by various fluorescent probes, such as fluo-3 and indo-1.

Channel: *Channel* is the term by which a flow cytometer characterizes the intensity of the signals emitted by a particle. Most cytometers divide the intensity of light signals into either 256 or 1024 channels. Signals with high channel numbers are brighter than signals with low channel numbers. However, the quantitative relationship between signals defined by one channel number and those defined by another will depend on the amplifier and photodetector voltage characteristics of a given protocol. Assignment of intensity to channel can be linear or logarithmic.

Coaxial flow: Coaxial flow is flow of a narrow core of liquid within the center of a wider stream. This type of flow is important in flow cytometry because it provides a means by which particles flowing through a relatively wide nozzle can be tightly confined in space, allowing accurate and stable illumination as they pass one by one through a light beam.

Coherence: Coherence is the property of light emitted from a laser such that it is remarkably uniform in color, polarization, and spatial direction. Spatial coherence allows a laser beam to maintain brightness and narrow width over a great distance.

Coincidence: Coincidence, in flow cytometry, is the appearance of two cells or particles in the laser beam at the same time. The flow cytometer will register these two cells as a single event (with approximately twice the fluorescence intensity and light scatter as a single cell). To avoid coincidence, the concentration of cells in the sample should be low, the laser beam should be small in the direction of flow, and the sample core should be narrow.

Compensation: Compensation is the way a flow cytometrist corrects for the overlap between the fluorescence spectra of different fluorochromes. Without compensation, fluorescence from a given fluorochrome may be included to some extent in the intensity value coming from a photodetector assigned to the detection of a different fluorochrome. Compensation can be electronic or can be applied by software during analysis of stored data.

Contour plot: A contour plot is one method for displaying data correlating two cytometric parameters. The density of particles at any place on the plot is used to generate contour lines (much as contour lines on a topographic map are used to describe the height of mountains at different points in the landscape). The particular algorithm used to generate the contour lines can make a great difference in the visual impact of the display.

Control: A control is a type of sample that, most particularly in flow cytometry, you generally have one fewer than you actually need.

Core stream: The core stream is the stream-within-a-stream that has been injected into the center of the sheath stream and is maintained there by the hydrodynamic considerations of laminar flow at increasing velocity. The core contains the sample particles that are to be analyzed in the flow cytometer. If the sample is injected too rapidly, the core stream widens and particles may be unequally illuminated. In addition, with a wide core stream, coincidence events are more likely.

Coulter volume: A cell's Coulter volume is the increased impedance that occurs as the cell displaces electrolyte when it flows through a narrow orifice. This increase in impedance is related to the volume of the cell.

Cross-match: Cross-matching is the process of testing the cells of a prospective organ donor with the serum of a prospective organ recipient for compatibility. The flow cytometric cross-match determines whether serum from the recipient contains antibodies that bind to the donor cells. Such binding constitutes a positive cross-match and is a contraindication to transplantation in that particular donor/recipient combination.

Cross-talk: Cross-talk is the signal from the "wrong" photodetector that results because the fluorescent light emitted by one fluorochrome contains some light of a wavelength that gets through the filters on a photodetector that is nominally specific for the fluorescence from a different fluorochrome. See **Compensation**.

CV: The coefficient of variation (CV) is defined as the standard deviation of a series of values divided by the mean of those values (expressed as a percentage). It is used in flow cytometry to describe the width of a histogram peak. Whereas in some proto-

cols it can be used to assess the variation in particle characteristics within a population, in DNA analysis (where all normal particles are assumed to have identical characteristics) it is
frequently used to assess the alignment of a flow cytometer (and
the skill of its operator). Discussions about CV have been known
to bring out rather primitive competitive instincts within groups
of flow cytometrists, who are usually friendly, well-adjusted
people.

Cytofluorimeter: *Cytofluorimeter* is a term synonymous with *flow
cytometer,* but with slightly antiquated overtones.

Deflection plates: Deflection plates are the two pieces of metal that
carry a high voltage to attract charged drops to one side or the
other of the main stream. It is by means of the charge on the
deflection plates that sorting of particles occurs. The deflection
plates also provide a good reason for flow cytometry operators
to keep their hands dry and wear rubber-soled shoes.

Detergent: Detergents, such as saponin, Triton X-100, or NP40, are
used to aid in the permeabilization of cell membranes in order
to facilitate staining of intracellular proteins.

Dichroic mirror: A mirror that by virtue of its coating reflects light
of certain wavelengths and transmits light of other wavelengths
is a dichroic mirror. Such mirrors are commonly used in flow
cytometry at 45° to the direction of incident light in order to
transmit light from specific fluorochromes to a particular photodetector and reflect light from different fluorochromes toward
different detectors. The exact angle at which the mirror is positioned affects its wavelength specificity.

Dot plot: A dot plot is a two-dimensional diagram correlating the
intensities of two flow cytometric parameters for each particle.
Dot plots suffer, graphically, from black-out in that an area of a
display can get no darker than completely black. If the number
of particles at a given point is very dense, their visual impact, in
comparison with less dense areas, will decrease as greater numbers of particles are displayed. Contour plots display the same
kind of correlation as dot plots but, because the levels of the
lines can be altered, can provide more visual information about

the density of particles at any given point in the correlation display. Dot plots can also be called *two-dimensional histograms*.

Drop delay: In flow sorting, the drop delay is the time between the measurement of the signals from a particle and the moment when that particle is just about to be trapped in a drop that has broken off from the main column of a vibrating stream. It is at this precise moment that the column of the stream must be charged (and then shortly thereafter grounded) if the drop containing a particle of interest is to be correctly charged so that it can be deflected to the left or right out of the main stream.

Emission: Emission is the loss of energy from an excited atom or molecule in the form of light. Although a long-lived form of light emission (known as *phosphorescence*) does occur, in flow cytometry the light emission with which we are primarily concerned occurs rapidly after excitation and is called *fluorescence*.

Erythrocyte: The red, hemoglobin-containing cells that occur in the peripheral circulation and are responsible for transporting oxygen are erythrocytes. Despite this vital physiological function, immunologists are apt to feel a certain amount of hostility toward erythrocytes because they outnumber white cells by about 1000 to 1 and therefore make flow analysis of white cells difficult unless they are removed by either lysis or centrifugation. Fetal erythrocytes are, conversely, of great interest when they appear in the maternal circulation because they can be used for prenatal screening and for diagnosis of fetal–maternal hemorrhage. When fixed in glutaraldehyde, chicken or trout erythrocytes fluoresce brightly at a broad range of wavelengths and are therefore a useful tool for aligning the cytometer beam; when fixed in ethanol, they are a useful DNA standard (containing less DNA than a normal human cell).

Euploid: Flow cytometrists use the term *euploid* to refer to a cell with the "correct" amount of DNA for its species. Because its sensitivity is limited and because it works only with total DNA content, a flow cytometer may not agree with a microscopist (who looks at chromosomes) in the classification of cells by this criterion. See also **Aneuploid**.

Event: An event is the name given by flow cytometrists to what most people would call a cell. A flow cytometer associates all light signals that occur without a gap in time with a single event and stores the intensities of the light in association with that event in the data file. If cells (or other particles) are spaced appropriately in the core stream and do not coincide in the laser beam, then an event is the same as a cell or particle. If cells do coincide in the laser beam, then an event may be two or more cells.

FACS: FACS is an acronym for *Fluorescence-Activated Cell Sorter*. It is a term coined by the Herzenbergs at Stanford University and used by Becton Dickinson for its instruments, but has come to be used generally as a term that refers to all instruments that analyze the light signals from particles flowing in a stream past a light beam. The term *flow cytometer* is perhaps more correct because it has neither the trademark connotations nor any reference to a sorting function, which most of these instruments no longer possess. The term *cytofluorimeter* is also used, but generally has antiquated overtones.

FCS format: *Flow Cytometry Standard* is a file format for flow cytometric data storage. Adherence to this format facilitates the programming for analysis of results. Manufacturers of cytometers have finally begun to embrace this standard.

Fetal hemoglobin: Fetal hemoglobin (HbF) is a form of hemoglobin that exists primarily in the erythrocytes of the fetus before birth. Because it has higher oxygen affinity than adult hemoglobin, it aids in the transfer of oxygen from the maternal to the fetal circulation. Antibodies against HbF can be used to detect fetal erythrocytes. By staining for fetal erythrocytes in the maternal peripheral blood, flow cytometry can be used to diagnose fetal–maternal bleeding across the placenta and also (in conjunction with DNA staining) to sort rare nucleated fetal erythrocytes for prenatal diagnosis of disease.

Filter (1): A glass filter (either colored glass or interference) modifies the light that passes through it. A neutral density filter reduces the intensity of a light beam without affecting its color. Short-pass, long-pass, and band-pass filters selectively alter the color of a light beam by transmitting light of restricted wavelength. If positioned at a 45° angle to the incident light, these filters are called *dichroic mirrors* (see above). The specifications of a band-pass filter are given in terms of the wavelength of the maximally transmitted light and the band width (in nm) at half-maximal intensity.

Filter (2): A paper or nylon filter removes particles above a certain size from liquid. In flow cytometry, these filters are important for cutting down on background noise by removing small particles (<0.22 μm) from the sheath fluid and also for removing clumps from your sample (<35 μm).

FITC: See Fluorescein.

Fixation: Fixation is the process by which the protein of cells is denatured, or cross-linked, and preserved. Fixation in flow cytometry is used to inactivate hazardous biological material and also to preserve stained cells when there is not immediate access to a flow cytometer. Fixation is also important in preserving proteins before detergent permeabilization for intracellular staining. Formaldehyde is often the fixative of choice for flow cytometry because it preserves the forward and side scatter characteristics of cells (but does cause some increase in their autofluorescence).

Flourescence: A common misspelling of *fluorescence* often found in keyword listings for articles on flow cytometry.

Flow: *Flow* is a term that has come to refer, colloquially, to the general technique of flow cytometry. For example "Do you use flow to analyze your cells?" or "Flow has revolutionized the study of immunology" or "I'm off to do flow."

Flow cell: The flow cell is the device in a flow cytometer that delivers the sample stream to the center of the sheath stream and then accelerates the flow velocity to maintain the cells spaced out from each other, in a narrowing core. In some cytometric configurations, the laser illuminates the stream within the flow cell; in other configurations the illumination occurs "in air" after the stream has left the flow cell. In the latter case, the term *nozzle* is more apt to be used. In a sorting instrument, the flow cell vibrates in order to allow drop formation. It also provides the electrical connections for charging and then grounding the stream at appropriate times.

Flow cytometry: See Chapters 1 through 12 of this book.

Flower: *Flower* is a term of affection in the Northeast of England—and one that could, with a slight change of pronunciation, be adapted to refer to a person who works in the field of flow cytometry.

Fluorescein: Fluorescein is a fluorescent dye that can be readily linked to proteins and that is therefore useful, when conjugated to specific antibodies, for lighting up cells with particular phenotypes. It is sometimes abbreviated as FITC because fluorescein isothiocyanate is the chemically active form of the molecule that is used in the conjugation process.

Fluorescence: Fluorescence is a form of light emitted by atoms or molecules when electrons fall from excited electronic energy levels to their lower, less-energetic ground state. *Fluorescence* is often misspelled as *flourescence*, so it pays to check both spellings when doing a key-word literature search.

Fluorochrome: A fluorochrome is a dye that absorbs light and then emits light of a different color (always of a longer wavelength). Fluorescein, propidium iodide, and phycoerythrin, for example, are three fluorochromes in common use in flow cytometry. *Fluorophore* is an equivalent term.

Fluorogenic substrate: A fluorogenic substrate is a nonfluorescent chemical that is an enzyme substrate and that yields a fluorescent product when processed by the enzyme. A fluorogenic substrate can therefore, by virtue of its increasing fluorescence, be used in a flow cytometer to measure the activity of a given enzyme within a cell. Fluorescein digalactopyranoside (FDG) is a fluorogenic substrate for the enzyme β-galactosidase.

Forward scatter: Forward scatter is light from the illuminating beam that has been effectively bent (refracted or otherwise deflected) to a small angle as it passes through a cell so as to diverge from the original direction of that beam. The intensity of the light bent to a small angle from the illuminating beam is related to the refractive index of the cell as well as to its cross-sectional area. The forward scatter signal causes a great deal of confusion because some people who call it a *volume* signal actually begin to think that it is closely correlated with a cell's volume. *Forward scatter* is often abbreviated as FSC or as FALS (for "forward angle light scatter").

G0/G1: In flow cytometry, the term refers to cells that have the 2C (diploid) amount of DNA and are therefore either not cycling (G0) or have just completed cytokinesis and have not yet begun again to make more DNA in preparation for a new cycle (G1).

G2/M: In flow cytometry, the term refers to cells that have the 4C (tetraploid) amount of DNA and are therefore either just completing DNA synthesis in preparation for mitosis (G2) or are involved in the steps of mitosis before cytokinesis (M).

Gain: Gain is a measure of the amplification applied to a signal. In flow cytometry, it occurs between the time of the initial detection of a light pulse and the final output that is stored as the channel on an ADC (and then in a data file). For a weak light pulse, the gain can be increased by the use of an amplifier, increasing the electrical signal after it leaves the photodetector; or it can be increased by altering the voltage directly applied to a photomultiplier tube.

Gate: Gate is a restriction placed on the flow cytometric data that will be included in subsequent analysis. A *live gate* restricts the data that will be accepted by a computer for storage; an *analysis gate* excludes certain stored data from a particular analysis procedure; a *sort gate* defines the cells that will be selected for sorting. See also **Back-gating**.

GFP: Green fluorescent protein comes from the jellyfish *Aequorea victoria*. It fluoresces green, which is of puzzling utility to the jellyfish but of great use to scientists because it can act as a reporter molecule or can be used to make fusion proteins visible. Outdoing the jellyfish, molecular biologists have created GFP analogs that fluoresce in different colors (BFP, CFP, and YFP).

Granularity: *Granularity* is a term sometimes used synonymously with *side scatter* to describe the light that is deflected to a right angle from the illuminating beam in a flow cytometer. The intensity of this light is related, in an imprecise way, to internal or surface irregularities of the particles flowing through the beam.

Granulocyte: Granulocytes are the major constituents of a class of white blood cells (called *polymorphonuclear cells*) that possess irregular nuclei and cytoplasmic granules and therefore have bright side scatter signals.

Hard drive: A hard drive is a storage medium for software and for data. Hard drives have relatively large memory capacity but, when not removable, require further back-up facilities because they will always become filled faster than you imagine.

Heath Robinson device: W. Heath Robinson (1872–1944) was an English humorous artist. A *Heath Robinson device* is the English term for a complex instrument that looks as if it had been designed by a committee. A Heath Robinson device usually works well but may be held together by chewing gum and string. In the context of state-of-the art flow cytometry, this term requires no further explanation. See also **Rube Goldberg device**.

High-speed sorting: High-speed sorting is a flow cytometric technique whereby adaptations to pressure and fluid controls allow high stream velocities and rapid drop generation. This decreases the probability of multiple cells in a single drop and consequently increases cell-recovery efficiency when sorting at fast rates. Such adaptations are important when large numbers of relatively rare particles are required.

Histogram: A one-dimensional histogram displays data from one parameter of a flow cytometric data file at a time.

Hydrodynamic focusing: Hydrodynamic focusing is the property of laminar flow in a stream of increasing velocity that maintains particles in the narrowing central core of a column of fluid.

Immunophenotyping: Immunophenotyping is the classification of normal or abnormal white blood cells according to their multiparameter surface antigen characteristics.

Interrogation point: *Interrogation point* is synonymous with *analysis point*.

Laser: Laser is an acronym for *L*ight *A*mplification by *S*timulated *E*mission of *R*adiation. Lasers are important in flow cytometry because, as a result of their coherent output, they are a means of illuminating cells with a compact, intense light beam that will produce fluorescence signals that are as bright as possible over a short time period.

Lens: A lens is a means of changing the shape of a beam of light. In flow cytometry, lenses are used to narrow the laser beam to a small profile at the stream. Some lenses produce a beam with a circular cross-sectional shape; others produce beams with an elliptical configuration. Lenses are also used in a flow cytometer to collect scattered light and fluorescence and then to transmit them to an appropriate photodetector.

Linear amplifier: A linear amplifier is one means of increasing the signal from a photomultiplier tube to make it stronger. A linear amplifier increases the signal in such a way that the output voltage from the amplifier is directly proportional to the input current derived from the photodetector. Linear amplifiers are used, in particular, for signals from the DNA of cells because the range of intensities being studied is small. See also **Logarithmic amplifier**.

List mode: List mode is a method of data storage in which the intensities of the signals from each photodetector that are generated by a single particle are stored in association with each other so that they act as a total flow cytometric description of that particle. A list mode data file consists of a long list of numbers, describing each cell in the order in which it passed the laser beam. The descriptive numbers from each particle can be correlated with each other (and with the descriptive signals from all other particles) in all possible ways. If you really want to understand the meaning of "list mode," you should ask Howard Shapiro to sing his song of that name.

Logarithmic amplifier: Logarithmic amplification is one means of modifying the signal from a photomultiplier tube. A logarithmic amplifier modifies the signal in such a way that the output voltage from the amplifier is in proportion to the logarithm of the input current derived from the photodetector. Logarithmic amplifiers are conventionally used for fluorescence signals from cellular proteins, where the range of signals (covering stained and unstained particles) is large. See **Linear amplifier** for comparison.

Lymphocyte: A lymphocyte is a particular type of white blood cell that is involved in many of an organism's immune responses. Subpopulations of lymphocytes with microscopically identical anatomy can be distinguished because their surface membranes contain different arrays of proteins. The staining of these proteins with fluorescently tagged monoclonal antibodies allows the subpopulations to be enumerated by flow cytometry.

Marker: *Marker* is a term that is often used to refer to a dividing line applied to a fluorescence intensity histogram in order to dichotomize particles into those that are to be called positively stained from those to be called unstained. Somewhat confusingly, it is also used by immunologists to refer to a significant protein on the surface of a cell.

Mean: The mean is a value that can be used to describe the fluorescence intensity of a population of cells. The arithmetic mean is calculated by adding up the intensities of n cells and then dividing that sum by n. It can be skewed considerably by the presence of a few very bright or very dim cells. The geometric mean is calculated by multiplying the intensities of n cells and then deriving the nth root of that product. It is less affected by outliers, but, accordingly, will not compare well with biochemical analysis. The calculation of a mean from a flow histogram will incorrectly evaluate any particles that lie in the highest and lowest (i.e., off scale) channels. See **Median** and **Mode**.

Median: Median is a value that can be used to describe the fluorescence intensity of a population of cells. If the cells were lined up in order of increasing intensity, the median value would simply be the channel number or intensity of the cell that is at the midpoint in the sequence. The median channel, because it is unaffected by off-scale events and outliers, is considered by many to be the best way to describe the intensity of a population. See **Mean** and **Mode**.

Mode: Mode is a value that can be used to describe the fluorescence intensity of a population of cells. The mode intensity is the channel number that describes the largest number of cells in a sample. Mode values are apt to be variable when intensity distributions are broad or when few particles have been analyzed. See **Mean** and **Median**.

Monocyte: Monocytes are a class of white blood cells that co-purify with lymphocytes in commonly used density gradient procedures. They tend to be promiscuously sticky for the nonspecific (Fc) ends of monoclonal antibodies and therefore can lead to misleading results in analysis of leukocyte subpopulations unless their Fc receptors are blocked in the staining procedure. Monocytes differ from lymphocytes in their forward and side scatter characteristics.

Necrosis: Necrosis is a cell's response to overwhelming injury. It has been referred to as "accidental cell death" by way of comparison with **apoptosis**, which is considered to be a more orderly and beneficial process. The flow cytometric hallmark of necrosis is permeabilization of the outer membrane (leading to propidium iodide fluorescence).

Nozzle: See **Flow cell**.

Obscuration bar: A forward scatter obscuration bar is a strip of metal or other material that serves to block out direct light from the illuminating beam. Any light reaching a photodetector will therefore be light that has been deflected around the bar by the physical characteristics of a particle interacting with the light. The width of the bar will define the narrowest angle by which the light must be deflected in order to reach the detector. In jet-in-air systems, there is also a side scatter obscuration bar, which decreases the background noise in the side scatter channel from laser light deflecting to wide angles from the stream–air interface.

Observation point: See **Analysis point**.

Optical bench: An optical bench is the stable table ;) that keeps the light beam, fluid streams, lenses, and photodetectors of a flow cytometer all precisely aligned with each other. Lack of stability in these components leads to artifactual results.

Optical sorting: Optical sorting is a flow method for deactivating cells or chromosomes by use of an additional (zapper) laser in conjunction with phototoxic dyes. Because the zapping laser can be deflected and returned to the flow stream at a rapid rate, optical sorting has the potential for being more rapid than traditional or high-speed droplet sorting.

Orifice: The orifice is the exit opening in a flow cell or nozzle. Its diameter can, in general, range from 50 to 300 μm. Cells or aggregates of cells anywhere near as large as the orifice will block it. The diameter of the orifice affects the possible drop frequencies for flow sorting.

Parameter: *Parameter* is the term applied to the types of information derived from a cell as it goes through the flow cytometer. The number of parameters measured by a cytometer is determined by the number of photodetectors present and also by any processing of the signals from each photodetector to provide "derived" parameters like signal area or signal width. Modest cytometers measure three parameters. Immodest cytometers measure 13 parameters. Average cytometers measure between four and six parameters. In the future, our standards may increase.

Particle: Particles are the objects that flow through flow cytometers. It is a general term that includes cells and chromosomes and beads and DNA fragments and gel microdrops, among others.

Peak reflect: The peak reflect method is an algorithm for analyzing DNA histograms to determine the number of cells in the S phase of the cell cycle. In the peak reflect method, the shapes of the 2C and 4C peaks are assumed to be symmetrical, thus allowing subtraction of the contribution from these two peaks from the S-phase cells between them.

Phase sensitivity: A phase-sensitive flow cytometer quantifies the life time of the fluorescence emitted by particles. The decay time of fluorescence from a given fluorochrome is altered by changing chemical environments, and therefore measurement of fluorescence lifetimes can provide information about the microenvironment surrounding the fluorescent probe.

Photodetector: A photodetector is a device that senses light and converts the energy from that light into an electrical signal. Within the operating range of the detector, the intensity of the electrical signal is proportional to the intensity of the light. Photomultiplier tubes and photodiodes are two types of photodetectors.

Photodiode: A photodiode is a type of photodetector used to detect relatively intense light signals. It does not have a high voltage applied to increase the current flow at its anode (output) end.

Photomultiplier tube: A photomultiplier tube is a type of photodetector used to detect relatively weak light signals. Its output current is increased by means of high voltage applied to it.

Phycoerythrin: Phycoerythrin is a fluorochrome derived from red sea algae. It is particularly useful in flow cytometric applications requiring dual-color analysis because, like fluorescein, it absorbs 488 nm light from an argon laser. However, it has a longer Stokes shift than fluorescein, and therefore the fluorescences of the two fluorochromes can be distinguished.

Plankton: Plankton are the small organisms that float in lakes and oceans and are moved passively, drifting with the tides and currents. Small plankton are excellent objects for flow cytometric analysis.

Precursor frequency: The precursor frequency of cells that have divided in response to a stimulus is the proportion of cells present at the time of the addition of the stimulus that were destined to begin division. It is a measure of the potential of a mixed population of cells for response to any particular activator and can be determined, by flow cytometry, with the use of fluorescent tracking dyes that stain cells with stability but dilute by half each time a cell divides.

Probe: *Probe* is a general term used, in flow cytometry, to refer to any chemical that fluoresces when it reacts or complexes with a specific class of molecules and therefore can be used to assay that molecule quantitatively. Propidium iodide and acridine orange are probes for nucleic acid because they complex specifically with nucleic acids and fluoresce brightly when they have reacted in this way. Fluo-3 is a calcium probe because it chelates calcium ions and fluoresces brightly when it is complexed with this ion.

Propidium iodide: Propidium iodide is a probe that can be used to measure quantitatively the amount of double-stranded nucleic acid that is present in a cell. After treatment of cells with RNase, it will measure the amount of DNA present. Because it does not cross an intact cell membrane, cells need to be treated with detergent or ethanol before it can be used to determine their DNA content. It can also be used to assess the viability of cells.

Quadrant: For dual-parameter analysis, a two-dimensional plot of particles according to the correlated fluorescence intensities of two colors is, traditionally, divided into four quadrants. The division is based on the background fluorescence of the unstained control sample. The quadrants, from this division, will contain (1) cells that have stained with the first fluorochrome only; (2) cells that have stained with both fluorochromes; (3) cells that remain unstained; and (4) cells that have stained with the second fluorochrome only. In other words, the quadrants are defined so that they delineate the two types of single positive cells (1 and 4), double-negative cells (3), and double-positive cells (2).

Rare event: Defining a rare event in flow cytometry is a bit like answering the question "How long is a piece of string?" Particles that are less than 0.1% of the total are considered somewhat rare and require some care for enumeration in a flow cytometer. Really rare particles are less than 0.001% of the total and can be detected, counted, and sorted with careful gating based on multiparameters.

Region: A region is a description, using flow cytometric characteristics, that will define a cluster of cells. A region, in the old days, was always rectangular—with an upper and lower channel limit for each parameter involved in the description. Now, a region can take any shape, based on the channel numbers defining all the vertices of an irregular polygon or ellipse. Multiple regions can be combined by Boolean logic (AND, OR, or NOT) to define a gate that is used for cell analysis. See **Gate**.

Rube Goldberg device: Reuben Lucius Goldberg (1883–1970) was an American humorous artist. A *Rube Goldberg device* is the American synonym for a Heath Robinson (see above) device. Mr. Goldberg and Mr. Robinson may or may not have known each other, but surely would have enjoyed each other's company.

S-FIT: An S-FIT approximation is a mathematical algorithm for guessing at which cells in a DNA-content histogram are actually in the S phase of the cell cycle. The S-FIT algorithm bases this guess on the shape of the DNA histogram in the middle region between the G0/G1 and the G2/M peaks.

S phase: S phase is the period of the cell cycle during which cells are in the process of synthesizing DNA in preparation for cell division. During S phase, cells have between the 2C amount of DNA normal to their species and the 4C amount of DNA, which is exactly double the 2C amount. It is the overlap of fluorescence intensity between cells in S phase and some of the cells with the 2C and 4C amounts of DNA that leads to uncertainty in the flow cytometric estimation of the S-phase fraction.

Sheath: Sheath is the fluid within which the central sample core is contained during coaxial flow within and from the flow cell of a flow cytometer.

Side scatter: Side scatter is light of the same color as the illuminating beam that bounces off particles in that beam and is deflected to the side. The "side" is usually defined by a lens at right angle (orthogonal orientation) to the line of the laser beam. Side scatter light (SSC) may alternatively be called *right-angle light scatter* (RALS) or *90° LS*. The intensity of this light scattered to the side is related in a general way to the roughness or irregularity or granularity of the surface or internal constituents of a particle.

Signature: *Signature* is a term used by marine flow cytometrists to refer to the flow cytometric characteristics of particular planktonic species in a sample of water. The signature of plankton is related, in particular, to the autofluorescent pigments that are present. *Signature* has also been used to describe the distinguishing histogram profile that results from the DNA-fragment analysis of different bacterial species.

SOBR: SOBR is a mathematical algorithm for guessing at which cells in a DNA-content histogram are actually in the S phase of the cell cycle. The SOBR (sum of broadened rectangles) algorithm bases this guess on the use of Gaussian-broadened rectangular distributions to attempt to fit the shape of the DNA fluorescence intensity histogram of the cells in question.

Stokes shift: The Stokes shift is the difference in color (expressed as either energy or wavelength) between the light absorbed by a fluorochrome and the light emitted when that fluorochrome fluoresces.

Tandem dye: Tandem dyes are conjugates of two different fluorochromes. They use resonance energy transfer to pass the light energy absorbed by the primary fluorochrome to the secondary fluorochrome, which then emits that energy as fluorescence. Tandem dyes occur naturally (e.g., the peridinin:chlorophyll complex or the aequorin:GFP complex) and can also be manufactured by organic chemists (e.g., PE-Cy5). In flow cytometry, tandem dyes are useful because they effectively increase the Stokes shift of a fluorochrome and thereby allow more fluorescence parameters to be observed with a single excitation wavelength.

Tetraploid: *Tetraploid* is a term used to describe cells with double the amount of DNA normal for a particular species. Malignant cells are frequently of the tetraploid type. Distinguishing this type of ploidy from normal G2 or M cells or, indeed, from clumps of two cells can be difficult with flow cytometry.

Threshold: The threshold is an electronic device by which an ADC can be made to ignore signals below a certain intensity. A forward scatter threshold is most commonly used in flow cytometry to exclude very small particles, debris, and electronic or optical noise from acquisition into a data file.

Time: Time is a parameter that is being used more frequently in conjunction with flow cytometric analysis. It is being used in relation to the functional analysis of cells for the rate of reaction (e.g., calcium ion flux or changes in membrane potential) in response to various stimuli. It is also being used, on a finer scale, in phase-sensitive measurements of fluorescence decay.

TUNEL: The TUNEL assay represents a historical low point in attempts to coin acronyms. The *T*erminal deoxynucleotidyl transferase–mediated d*U*TP *N*ick *E*nd-*L*abeling assay labels the ends of DNA with fluorescent UTP. Apoptotic cells often have fragmented DNA; these fragments will provide more substrate for the enzyme terminal deoxynucleotidyl transferase. Apoptotic cells can therefore be enumerated by flow cytometry according to their increased intensity in the TUNEL assay.

Volume: Volume is a useful and definite characteristic of any particle —but one that is not easily amenable to flow cytometric analysis. "Coulter volume" does measure volume to a close approximation, but "forward scatter" does not.

Wavelength (1): Wavelength is a characteristic of light that is related exactly to its energy content and also (with light to which our eyes are sensitive) to its color. Light of short wavelength lies toward the bluer region of the spectrum and has more energy than light of longer wavelength. Wavelength is used to describe the characteristics of filters, of dichroic mirrors, of laser beams, and of the absorption and emission of fluorochromes.

Wavelength (2): In sorting flow cytometry, *wavelength* is a term used to describe the distance between drops as they form from a vibrating fluid jet. This drop wavelength is determined by the diameter of the stream as well as by its velocity.

Figure Credits

Fig. 1.1. Reprinted from Alberts B, et al. (1989). Molecular Biology of the Cell, 2nd edition. New York: Garland Publishing.

Fig. 1.2. Reprinted from Kamentsky LA and Melamed MR (1967). Spectrophotometric cell sorter. *Science* **156**:1364–1365. © 1967 by the American Association for the Advancement of Science.

Fig. 1.3. Reprinted, with permission, from the Lawrence Livermore National Laboratory, operated by the University of California under contract to the U.S. Department of Energy.

Fig. 1.4. Photograph by Edward Souza. Reprinted, with permission, from the Stanford University News Service.

Fig. 1.5. Photographs reprinted from Beckman Coulter, Inc., Miami, FL; BD Biosciences, San Jose, CA; and Dako A/S, Glostrup, Denmark.

Fig. 3.2. Reprinted from Givan AL (2001). Principles of flow cytometry: an overview. Darzynkiewicz Z, et al. (eds). Cytometry, 3rd edition. San Diego: Academic Press, pp 19–50.

Fig. 3.6. Reprinted from Pinkel D and Stovel R (1985). Flow chambers and sample handling. Van Dilla MA, et al. (eds). Flow Cytometry: Instrumentation and Data Analysis. London: Academic Press, pp 77–128.

Fig. 3.7. Adapted from BD Biosciences, San Jose, CA.

Fig. 3.10. Reprinted from Blakeslee A (1914). Corn and men. *J. Hered.* **5**:512.

Fig. 3.11. Reprinted from Givan AL (2001). Principles of flow cytometry: an overview. Darzynkiewicz Z, et al. (eds). Cytometry, 3rd edition. San Diego: Academic Press, pp 19–50.

Fig. 4.8. Reprinted with permission of John Wiley & Sons, Inc. © 1990 from Dean PN (1990). Data processing. Melamed MR, et al. (eds). Flow Cytometry and Sorting. New York: Wiley-Liss, pp 415–444.

Fig. 5.3. Reprinted from Spectra Physics Lasers, Inc., Mountain View, CA.

Fig. 5.6. Reprinted with permission of John Wiley & Sons, Inc. © 1995 from Shapiro HM (1995). Practical Flow Cytometry, 3rd edition. New York: Wiley-Liss.

Fig. 5.7. Reprinted from Givan AL (2001). Principles of flow cytometry: an overview. Darzynkiewicz Z, et al. (eds). Cytometry, 3rd edition. San Diego: Academic Press, pp 19–50.

Fig. 6.1. Reprinted from Kessel RG and Kardon RH (1979). Tissues and Organs: a Text-Atlas of Scanning Electron Microscopy. San Francisco: WH Freeman & Co.

Fig. 6.3. Printed with permission from Ian Brotherick.

Fig. 6.6. Reprinted from Givan AL (2001). Principles of flow cytometry: an overview. Darzynkiewicz Z, et al. (eds). Cytometry, 3rd edition. San Diego: Academic Press, pp 19–50.

Fig. 6.15. Reprinted with permission of John Wiley & Sons, Inc. © 2000 from Loken MR and Wells DA (2000). Normal antigen expression in hematopoiesis. Stewart CC and Nicholson JKA (eds). Immunophenotyping. New York: Wiley-Liss, pp 133–160.

Fig. 6.17. Reprinted from Horan PK, et al. (1986). Improved flow cytometric analysis of leucocyte subsets: simultaneous identification of five cell subsets using two-color immunofluorescence. *Proc. Natl. Acad. Sci.* **83**:8361–8363.

Fig. 7.1. Unpublished results from McNally A and Bauer KD, reprinted with permission from Bauer KD and Jacobberger JW (1994). Analysis of intracellular proteins. Darzynkiewicz Z, et al (eds). Flow Cytometry, 2nd edition. San Diego: Academic Press, pp 351–376.

Fig. 7.3. Reprinted (in modified form) with permission of John Wiley & Sons, Inc. © 1995 from Brotherick I, et al (1995). Use of the biotinylated antibody DAKO-ER 1D5 to measure oestrogen receptors on cytokeratin positive cells obtained from primary breast cancer cells. *Cytometry* **20**:74–80.

Fig. 8.3. Reprinted from Alberts B, et al. (1989). Molecular Biology of the Cell, 2nd edition. New York: Garland Publishing.

Fig. 8.4. Reprinted with permission of John Wiley & Sons, Inc. © 1990 from Gray JW, et al. (1990). Quantitative cell-cycle analysis. Melamed MR, et al. (eds). Flow Cytometry and Sorting. New York: Wiley-Liss, pp 445–467. The work was performed at the University of California Lawrence Livermore National Laboratory under the auspices of the U.S. Department of Energy.

Fig. 8.7. Reprinted (A,D) from Dean PN (1987). Data analysis in cell kinetics. Gray JW and Darzynkiewicz Z (eds). Techniques in Cell Cycle Analysis. Clifton, NJ: Humana Press, pp 207–253; and (B,C) from Dean PN (1985). Methods of data analysis in flow cytometry. Van Dilla MA, et al. (eds). Flow Cytometry: Instrumentation and Data Analysis. London: Academic Press, pp 195–221.

Fig. 8.9. Reprinted with permission of John Wiley & Sons, Inc. © 1989 from Peeters JCH, et al. (1989). Optical plankton analyser. *Cytometry* **10**:522–528.

Fig. 8.10. Reprinted in modified form with permission from Michael Ormerod.

Fig. 8.13. Reprinted with permission of Oxford University Press from McNally NJ and Wilson GD (1990). Measurement of tumour cell kinetics by the bromodeoxyuridine method. Ormerod MG (ed). Flow Cytometry: A Practical Approach. Oxford: IRL, pp 87–104.

Fig. 8.14. Reprinted with permission of John Wiley & Sons, Inc. © 1990 from Darzynkiewicz Z and Traganos F (1990). Multiparameter flow cytometry studies of the cell cycle. Melamed MR, et al. (eds). Flow Cytometry and Sorting. New York: Wiley-Liss, pp 469–501.

Fig. 8.15. Reprinted with permission of John Wiley & Sons, Inc. © 1990 from Darzynkiewicz Z (1990). Probing nuclear chromatin by flow cytometry. Melamed MR, et al. (eds). Flow Cytometry and Sorting. New York: Wiley-Liss, pp 315–340.

Fig. 8.17. Reprinted with permission of John Wiley & Sons, Inc. © 1998 from Juan G, et al. (1998). Histone H3 phosphorylation and expression of cyclins A and B1 measured in individual cells during their progression through G2 and mitosis. *Cytometry* **32**:71–77.

Fig. 8.18. Printed with permission from James Jacobberger.

Fig. 8.19. Reprinted with permission of John Wiley & Sons, Inc. © 1988 from Cram LS, et al. (1988). Overview of flow cytogenetics for clinical applications. *Cytometry [suppl]* **3**:94–100.

Fig. 8.20. Reprinted with permission of John Wiley & Sons, Inc. © 1990 from Gray JW and Cram LS (1990). Flow karyotyping and chromosome sorting. Melamed MR, et al. (eds). Flow Cytometry and Sorting. New York: Wiley-Liss, pp 503–529. The work was performed at the University of California Lawrence Livermore National Laboratory under the auspices of the U.S. Department of Energy.

Fig. 8.21. Reprinted with permission of John Wiley & Sons, Inc. © 1990 from Gray JW and Cram LS (1990). Flow karyotyping and chromosome sorting. Melamed MR, et al. (eds). Flow Cytometry and Sorting. New York: Wiley-Liss, pp 503–529. The work was performed at the University of California Lawrence Livermore National Laboratory under the auspices of the U.S. Department of Energy.

Fig. 9.1. Reprinted from Cytomation, Inc., Fort Collins, CO.

Fig. 10.1. Printed with permission from Ben Givan.

Fig. 10.2. Reprinted with permission of John Wiley & Sons, Inc. © 2000 from Loken MR and Wells DA (2000). Normal antigen expression in hematopoiesis. Stewart CC and Nicholson JKA (eds). Immunophenotyping. New York: Wiley-Liss, pp 133–160.

Fig. 10.3. Printed with permission from Carleton Stewart.

Fig. 10.4. Printed with permission from Léonie Walker.

Fig. 10.5. Unpublished results from Sutherland R, reprinted with permission of John Wiley & Sons, Inc. © 2000 from Gee AP and Lamb LS Jr (2000). Enumeration of CD34-positive hematopoietic progenitor cells. Stewart CC and Nicholson JKA (eds). Immunophenotyping. New York: Wiley-Liss, pp 291–319.

Fig. 10.6. From Davis BH et al (1998). Detection of fetal red cells in fetomaternal hemorrhage using a fetal hemoglobin monoclonal antibody by flow cytometry. Reprinted with permission from *Transfusion* **38**:749–756, published by the American Association of Blood Banks.

Fig. 10.7. Reprinted from Yuan J, Hennessy C, et al. (1991). Node negative breast cancer: the prognostic value of DNA ploidy for long-term survival. *British J. Surg.* **78**:844–848.

Fig. 11.1. Reprinted (with modifications) with permission of John Wiley & Sons, Inc. © 1995 from Shapiro HM (1995). Practical Flow Cytometry, 3rd edition. New York: Wiley-Liss.

Fig. 11.5. Reprinted from Veldhuis MJW and Kraay GW (2000). Application of flow cytometry in marine phytoplankton research: current applications and future perspectives. *Scientia Marina* **64**:121–134.

Fig. 11.6. Reprinted with permission of the Department of Fisheries and Oceans and the Minister of Supply and Services, Canada, 1991, from Chisholm S, et al. (1986). The individual cell in phytoplankton ecology. *Can. Bull. Fish. Aquat. Sci.* **214**:343–369.

Fig. 11.7. Printed with permission of Sandra Shumway.

Fig. 11.8. Reprinted with permission of YN Jan from Bier E, et al. (1989). Searching for pattern and mutation in the *Drosophila* genome with a P-*lac Z* vector. *Genes and Dev.* **3**:1273–1287.

Fig. 11.9. Printed with permission from Mark Krasnow.

Fig. 11.10. Reprinted with permission from Krasnow M, et al. (1991). Whole animal cell sorting of *Drosophila* embryos. *Science* **251**:81–85. © 1991 by the American Association for the Advancement of Science.

Fig. 11.11. Reprinted from Weaver JC (1990). Sampling: a critical problem in biosensing. *Med. Biol. Eng. Comput.* **28**:B3–B9.

Fig. 11.12. Reprinted from Weaver JC, et al. (1991). Rapid clonal growth measurements at the single cell level: gel microdroplets and flow cytometry. *Bio/Technology* **9**:873.

Fig. 11.13. Reprinted with permission of John Wiley & Sons, Inc. © 1990 from Van Dilla M, et al. (1990). Applications of flow cytometry and sorting to molecular genetics. Melamed MR, et al. (eds). Flow Cytometry and Sorting. New York: Wiley-Liss, pp 562–604.

Index

In this index, page numbers followed by the letter "f" designate figures; page numbers followed by "t" designate tables. Page numbers in **boldface** designate glossary definitions.

Absorption, **235**
  by electrons in atom, 60, 60f
  by fluorochromes, 66–72, 66f, 68f, 69t
Acquisition, 44, **235**
Acridine Orange, 124t, **236**
  Darzynkiewicz techniques, 143–144, 143f, 144f
Activation Markers, 196–197, **236**
ADC. *See* Analog-to-digital converter
Aequorea victoria, 209
Aerosols, 177, **236**
Aggregates/clumped cells
  blocking orifice, 25
  confused with large cells, 86
  confused with tetraploid cells, 131
  in cell cycle analysis, 137–139, 137f, 139f
  in gel microdrops, 212
  in sorting, 168
AIDS, 178, 181, 182f
Alexa dyes, 64t, 69t, 72
Algae, **236**
  aquatic studies of, 202–206
  fluorochromes from, 70–72
Algorithms for cell cycle analysis, 135–137, 136f, 139, **251, 254, 255**
Allophycocyanin, 64t, 68f, 69t, 72, 203
Amplification, **248**
  decades full scale, 34
  interaction with ADC, 35–37
  linear, for DNA, 126
  logarithmic vs linear, 31–37

Analog-to-digital converter (ADC), 35–37, 96–97, **236**
Analysis, **237**; *see* Data analysis
Analysis gate, 54–55, 55f
Analysis point, 21, 23f, 26, **237**
  in time delay calculation for sorting, 163
  photodetectors around, 28–30
Aneuploidy, 127–130, 127f, 141, 145, 187f, 188, 188f, **237**
Ankylosing spondylitis, 184
Annexin V, 151, 152f, 153f, **237**
Antibodies
  against platelets, 183
  against fetal hemoglobin, 185, 221
  anti-Rh, 220
  binding site calibration, 97–98
  conjugation with fluorochromes, 1, 67, 71
  controls, 90–94
  cytotoxic, in transplantation, 190
  direct staining with, 87–88
  effect on sorted cells, 173
  for gating, 108–113
  indirect staining with, 88–90, 89f
  intracellular staining with, 115–118
  leukocyte analysis and, 100
  surface staining with, 87–90
  specificity of, 91
  to distinguish CD antigens, 87
  see also Monoclonal antibodies
Antigens
  binding to antibodies, 87–88